Depiction and Designing of Plant Genomes

Series Editor
Chittaranjan Kole and Phullara Kole

The Cannabis Genome
A New Era in Genetics, Genomics, and Breeding
Edited by Angel Fernandez i Marti and Jaime Prohens

Designing Sorghum Genome for NextGen Agriculture
Edited by Dinesh Kumar Saini and S. V. Krishna Jagadish

The Turmeric Genome
Edited by Prasath Duraisamy, Aarthi Sounderarajan, and Chittarajan Kole

The Aloe Genome
Edited by Vinod Chhokar and Alka Jangra

The Melon Genome
Designing for NextGen Agriculture
Edited by Rakesh Kumar and Chittaranjan Kole

For more information, please visit our website: Depiction and Designing of Plant Genomes – Book Series – Routledge & CRC Press

The Melon Genome

Melon is a major cucurbit crop cultivated on 1.09 million hectares, producing 29.5 million tons. It is grown on nearly every continent, highlighting its significant economic value. Both the public and private sectors operate dedicated melon breeding programs focused on crop improvement. This book serves as an essential resource for breeders, geneticists, students, and industry professionals, providing the most up-to-date information to guide melon crop enhancement. Additionally, it also aims to unify the objectives of public and private researchers.

The Melon Genome: Designing for NextGen Agriculture presents the latest research as a foundation for future melon crop improvement. Without effective breeding programs, sustaining melon production becomes challenging. This book integrates various aspects of melon breeding, including genetic diversity, genomics, and germplasm resources, for improving yield, quality, and resistance to biotic stresses.

Key features:

- Wide biodiversity, classes, and molecular diversity exist in melon germplasm
- Utilization of advanced sequencing technologies in melon breeding
- Germplasm resources and strategies for improving the melon crop for various traits

Content for this book has been contributed by eminent scientists who have been engaged in melon research for many decades. This book will be welcomed by scientists aiming to develop the next generation of melon cultivars.

Rakesh Kumar is a globally renowned breeder at Syngenta with 16 years of experience in commercial breeding. The cultivars he has developed have had a worldwide impact and are grown on nearly every continent. His varieties have received the 'Most Innovative Product' award from Seed World and the 'Fruit Logistica Silver Innovation Award'.

Chittaranjan Kole is an internationally renowned academician with a spectacular professional career spanning over 40 years. His lifetime original scientific contributions, specifically in the fields of plant genomics and biotechnology, have been globally appreciated. His scholarly publications include 170 research articles, mostly published in peer-reviewed international journals, 35 book chapters, and over 200 books with globally reputed publishers. His academic contributions have been profusely appreciated by seven Nobel laureates. His research contributions have been recognized with several international awards and fellowships. Professor Kole has worked as a faculty member, head of department, dean, or vice chancellor in three universities in India and as a visiting professor in three universities in the United States. Currently, Professor Kole is acting as the founder president of Genome India International and of two other international organizations, including the International Climate Resilient Crop Genomics Consortium and the International Phytomedomics and Nutriomics Consortium.

The Melon Genome
Designing for NextGen Agriculture

Edited by
Rakesh Kumar and Chittaranjan Kole

CRC Press
Taylor & Francis Group
Boca Raton London New York

CRC Press is an imprint of the
Taylor & Francis Group, an **informa** business

Designed cover image: Cover design by Chittaranjan Kole; Image courtesy of Rakesh Kumar

First edition published 2026
by CRC Press
2385 NW Executive Center Drive, Suite 320, Boca Raton FL 33431

and by CRC Press
4 Park Square, Milton Park, Abingdon, Oxon, OX14 4RN

CRC Press is an imprint of Taylor & Francis Group, LLC

ISBN: 978-1-041-20244-8 (hbk)
ISBN: 978-1-041-20245-5 (pbk)
ISBN: 978-1-003-71575-7 (ebk)

DOI: 10.1201/9781003715757

Typeset in Times LT Std
by Apex CoVantage, LLC

This book series is dedicated to our brothers/brothers-in-law and sisters/sisters-in-law –

Bansi Badan Kole
Niranjan Kole
Bharati Nandi Nee Kole
Arati Kundu Nee Kole
Ram Ranjan Kole

Contents

Preface to the Series

Genome sequencing of the model plant *Arabidopsis* in 2000, followed by the crop plant rice in 2002, flagged genome sequencing in an array of plants. As of March 5, 2024, 4604 genomes of 1482 different plant species have been sequenced, and the number has continued to grow exponentially. The list includes several crop plants as well as some model plants. The sequencing of the nuclear genomes was followed by the sequencing of extranuclear genomes, such as mitochondria and chloroplast, and several studies on functional genomics. However, detailed reports on them are scattered in the literature, including research articles and reviews. Therefore, a comprehensive compilation of deliberations on the sequences with enumeration of the sequences, annotation, assembly, gene families and comparative genomics of these useful genomes and elucidation of functional genomics, including transcriptomics, metabolomics and proteomics in these higher plants is timely.

A number of higher plants provide the basic F5 – food, feed, fuel, fiber and furniture – needs and nutraceuticals, beverages, luxury and aesthetic items. In the last century, genetic improvement in these plants regarding their utilities was done mainly by traditional breeding. The advent of molecular breeding and genetic engineering in the last two decades of the last century strengthened the breeding strategies in the economic plants. Advances in plant genomics have ultimately revolutionized the concepts and strategies of crop improvement. The employment of high-throughput genotyping and phenotyping together with other genomics-assisted breeding tools like gene discovery, allele mining, genome-wide association studies, genomic selection and genome-wide prediction, among others, has facilitated accelerated and precise plant breeding. The recent emergence of the genome editing technique has enormously supplemented the molecular and genomics tools, making it possible to design and tailor the genomes according to need. However, success stories of next-generation breeding techniques need to be compiled for easy access to the readers.

This present book series, titled Depiction and Designing of Plant Genomes, includes a number of books, each dedicated to the deliberations on genomic attributes and/or genetic improvement of a higher plant, and hopefully will serve as handy compendium for readers.

The books of this series have been edited by highly reputed scientists, and we remain thankful for their invaluable academic efforts and cooperation. We are also thankful to the publisher and specifically to its Senior Commissioning Editor (Chemistry/Agricultural Sciences), Dr. Renu Upadhyay, and her Editorial Assistants, Jyotsna Jangra and Tisha Sharma, for their outstanding and timely help and cordial cooperation since the inception of this book series.

Finally, we must express our thanks to our children, Sourav and Devleena, for assisting us in editing the books of this series and bearing with us for our channeling the time that should have been kept for them toward editing this book series.

Kolkata, August 3, 2024

Chittaranjan Kole
Phullara Kole

Preface

The Melon Genome: Designing for NextGen Agriculture was written to guide melon breeding efforts in addressing future challenges imposed by climate change. Melon is an economically important crop that is cultivated worldwide. Stressors, including drought, salinity, flooding, pathogens, and insect pests, continuously challenge melon production. Although chemical strategies have been developed to manage certain obstacles, they do not represent a sustainable long-term solution. Instead, breeders must strive to establish climate-resilient cultivars with durable resistance to abiotic and biotic stresses. For decades, researchers and breeders have enhanced agronomic traits in melon through genetic improvement. However, few resources focus on the techniques and resources employed in melon breeding. This book lays the foundation for melon crop improvement, discussing industrial challenges and genomic resources for enhancing yield, fruit quality, and resistance to various plant pathogens. The material strives to assist breeders, trait scientists, molecular biologists, students, and anyone interested in melon research.

The expanse of commercial and public melon research programs has driven the evolution of breeding approaches and stands as a testament to the crop's importance. Traditional breeding techniques, such as pedigree breeding, single-seed descent, and backcross breeding, have long served as the basis for melon advancement. Since 2012, melon researchers have benefited from the availability of a reference genome that enables the physical mapping and annotation of quantitative trait loci intervals. Additionally, the availability of digital tools for high-throughput phenotyping and advancements in sequencing technologies have facilitated genome-based selection rather than marker-assisted trait selection. Through my years of work in melon research, I have recognized the importance of integrating past knowledge with emerging innovations to address future breeding challenges. The objective of this book is to provide researchers with the most up-to-date information to guide melon crop improvement.

Chapter 1 discusses global melon production and outlines the significant challenges impacting its cultivation. In particular, we focus on how climate change, as well as biotic and abiotic factors, impact melon yield and quality. Chapter 2 addresses the biodiversity, classification, and molecular diversity of melon. The crops' high genetic and phenotypic diversity serves as a vital resource for breeding. Chapter 3 explores recent genomic research and the impact of new sequencing technologies on melon breeding strategies. This comprehensive chapter examines breeding methods that target essential traits and the candidate genes associated with them. Chapter 4 discusses the influence of sex expression in melon breeding strategies. Cucurbits exhibit diverse patterns of sex expression, which play a role in fruit morphology and seed production. Finally, Chapters 5, 6, and 7 focus on how various strategies can be leveraged to enhance melon yield, fruit quality, and resistance to biotic stress.

This book represents the culmination of a year-long effort and encompasses contributions from distinguished scientists with decades of experience in melon research. We hope this serves as a valuable resource for breeders, geneticists, and students working in crop improvement. We wish to acknowledge all collaborators who generously contributed their time and expertise to the creation of this book.

Rakesh Kumar
Chittaranjan Kole

Contributors

Catherine Dogimont
INRAE Genetics and Improvement of Fruits and Vegetables
Montfavet, France

Venkata Rao Ganaparthi
Coastal Research and Education Center,
Clemson University Charleston South Carolina
USA

Rakesh Kumar
Syngenta Vegetables Seed
Woodland, California
USA

James D. McCreight
Sam Farr United States Crop Improvement and Protection Research Center
U.S. Department of Agriculture, Agricultural Research Service
Salinas, California
USA

Shaonpius Mondal
Department of Entomology
University of Nebraska–Lincoln
Lincoln, Nebraska
USA

Akito Nashiki
Agricultural Solutions
Chuo-ku, Tokyo
Japan

Nebahat Sari
Rutgers, The State University of New Jersey, Department of Plant Biology
New Brunswick, New Jersey
USA

Ilknur Solmaz
Department of Horticulture, Cukurova University
Sarıcam, Adana
Türkiye

Prabin Tamang
U.S. Department of Agriculture, Agricultural Research Service
Natural Products Utilization Research
Oxford, Mississippi
USA

Lingli Tang
Zhengzhou Fruit Research Institute, Chinese Academy of Agriculture Sciences
Zhengzhou, Henan
China

William M. Wintermantel
Sam Farr United States Crop Improvement and Protection Research Center
U.S. Department of Agriculture, Agricultural Research Service
Salinas, California
USA

Guangwei Zhao
Zhengzhou Fruit Research Institute, Chinese Academy of Agriculture Sciences
Zhengzhou, Henan
China

Abbreviations

ABPs	Actin-binding proteins
ADFs	Actin depolymerizing factors
AFLPs	Amplified fragment length polymorphisms
CNV	Copy number variation
CDM	Cucurbit downy mildew
CGMMV	Cucumber green mottle virus
CMV	Cucumber mosaic virus
CVYV	Cucumber vein yellowing virus
DM	Downy mildew
CmEIN2	Ethylene insensitive 2
eIFs	Eukaryotic translation initiation factors
GWAS	Genome-Wide Association Analysis
GBS	Genotyping-by-sequencing
ICuGI	International Cucurbit Genomics Initiative
LRRs	Leucine-rich repeat receptors
MAS	Marker-assisted selection
Mca	*Momordica charantia* L. var. *abbreviata Ser.*
MWMV	Moroccan watermelon mosaic virus
NGS	Next-generation sequencing
PPR	Pentatricopeptide repeat
PAV	Presence/Absence Variation
Pi	Phosphorus
K^+	Potassium
QTLs	Quantitative trait loci
RAPD	Random amplified polymorphic DNA
RGAs	Resistance gene analogs
SNPs	Single nucleotide polymorphisms
SSC	Soluble solids content
SV	Structural variation
SUP	SUPERMAN
SPS	Sucrose phosphate synthase
SS	Sucrose synthase
T2T	Telomere-to-telomere
TEs	transposable elements
VOCs	Volatile organic compounds
ZFP	Zinc finger protein
ZYMV	Zucchini yellow mosaic virus

1 Challenges Facing the Melon Industry

Rakesh Kumar

1.1 INTRODUCTION

Melon is a member of the Cucurbitaceae family, which also includes cucumbers, watermelons, squash, pumpkins, and gourds. The fruit crop is prized for its sweet taste, rich flavor, high nutritive value, and medicinal properties. Although cultivated worldwide, melon production is concentrated in temperate to tropical regions. The crop is highly diverse, with major horticultural groups including Cantaloupe, HoneyDew, Yellow Canary, Piel de Sapo, Charentais, Italo-American, Hami, and Ananas. This diversity generates numerous avenues for genetic improvement by leveraging beneficial traits across typologies.

Melons were first domesticated in Asia and Africa for their seeds, which are rich in protein and fats (Zeven and De Wet, 1982). Over time, other horticultural traits, such as flavor, aroma, and total soluble solids, were enhanced through breeding. Manchali et al. (2021) characterized diverse melon cultivars for their nutritional attributes, noting that the high content of carotenoids represents a rich source of vitamin A. In addition, melons contain sugars, protein, and antioxidants, including phenolic compounds and ascorbic acid (Chang et al., 2015; Gómez-García et al., 2020). Sugar content is a significant indicator of maturity and quality in melon. While glucose and fructose contribute to the sugar content of melons, sucrose is the primary source of the crop's sweetness (Schaffer et al., 1987; Burger and Schaffer, 2007).

Due to their extensive cultivation, melons are a valuable reservoir of genetic diversity, which can be leveraged to enhance crop resilience and promote sustainable agriculture. In 2023, 29.5 million tons of melon were produced on 1.09 million hectares of farmland (FAOSTAT, 2025). Production in Asian countries accounts for 75% of the total acreage, followed by the Americas (13%), Europe (6.5%), and Africa (4.4%). Cultivation is primarily centered in China (14 million tons), followed by India (1.4 million tons), and Türkiye (1.4 million tons). Other major production regions include Kazakhstan, Afghanistan, Guatemala, Iran, Spain, Brazil, and Italy (FAOSTAT, 2025). Each country has a specific product profile (typology) based on consumer demand and environment. For instance, cantaloupes are primarily grown in the Americas and Europe, Ananas in the Middle East, Piel de Sapo in Spain, yellow canary in Brazil, and Charentais in France.

Because melons are highly sensitive to temperature, day length, humidity, light, and abiotic and biotic stresses, they must be adapted to the agro-climatic region in which they are grown. Most breeding efforts target traits desired by growers, including

DOI: 10.1201/9781003715757-1

high yield, high sugars, field stability, flesh strength, dense netting (in cantaloupe), resistance to fruit cracking, and resistance to fungal and viral pathogens. Melon production follows seasonal migration patterns, and the fruit is often exported from one region to another. For example, most melons available in the United States during winter originate from Guatemala, Honduras, and Mexico, while Europe relies on imports from Brazil during the same season. In the summer, Spain exports melons to France, Germany, and the Netherlands, while Australia exports to Japan (FAOSTAT, 2025). The United States is the top importer of melon, followed by the Netherlands, France, the United Kingdom, and Canada. Spain is the top exporter, followed by Guatemala, Brazil, and Honduras (FAOSTAT, 2025). Melons bred for export require traits that enable the fruits to withstand long-distance shipping, such as high post-harvest shelf life, sugar stability, firm flesh, and resistance to rotting and discoloration. These melons must satisfy the trait requirements at every step in the value chain, which includes growers, shippers, retailers, and, ultimately, consumers.

Global melon production has increased steadily since the 1990s in terms of area harvested, yield, total production, and production per capita. According to the Food and Agriculture Organization, global melon production is projected to exceed 40 million tons annually. In parallel, the international market for melon-derived beverages is expected to surpass $3.5 billion, with a compound annual growth rate of 6% from 2021 to 2028 (Shahwar et al., 2024). This growth may be attributed to breeding efforts, improved agronomical practices, mechanization, and the efficacy of fungicides, insecticides, and herbicides. Modern advancements in molecular biology, omics, and phenotyping have accelerated melon breeding by enabling cost-effective and strategic breeding through tools such as quantitative trait locus (QTL) mapping, marker-assisted trait selection, and genomic selection. These advanced sequencing techniques facilitate the characterization of diverse melon accessions and support the identification of genes associated with valuable traits, such as productivity, abiotic and biotic resistance, shelf life, and fruit quality (Garcia-Mas et al., 2012; Zhang et al., 2019; Oren et al., 2022; Sun et al., 2022).

Despite these advancements, melon crop production remains vulnerable to numerous biotic and abiotic stresses. Climate change is one of the most significant threats to global food security, as it exacerbates factors known to impact crop yield, including drought, high salinity, and extreme temperatures (Shahwar et al., 2024). These yield-reducing effects are further compounded by deteriorating soil and water quality. Furthermore, melon cultivation is affected by a wide range of diseases and pests. Major diseases affecting melon include Fusarium wilt, powdery mildew, downy mildew, gummy stem blight, cucumber mosaic virus (CMV), cucurbit yellow stunting disorder virus (CYSDV), tomato leaf curl New Delhi virus (TocNDV), squash mosaic virus (SqMV), zucchini yellow mosaic virus (ZYMV), and watermelon mosaic virus (WMV) (Sharma et al., 2025; Tian et al., 2024). Pests such as fruit flies, nematodes, cucumber beetles, aphids, and thrips cause devastating losses in melon crops, both through direct damage and by transmitting diseases.

Although chemical interventions can mitigate several abiotic and biotic stresses, they do not represent a sustainable solution. Most production problems are highly localized, necessitating varieties adapted to local environmental conditions. While it may be a challenging feat, melon breeders have a responsibility to develop cultivars that resist relevant stressors. The polygenic nature of many key melon traits further complicates this task. Traits related to yield, total soluble solids, and resistance to diseases and pests are quantitatively inherited and, therefore, exhibit high genotype × environment (GE) interactions (Dogimont and Sari, 2022; Kopeć, 2024). Traits with strong GE interactions are extremely hard to breed for, especially under the rapidly changing conditions induced by climate change.

In the Americas, a significant emphasis has recently been placed on breeding melons with a long shelf life (LSL). This transition began in the mid-2000s, when efforts shifted away from highly aromatic and medium shelf-life cultivars. LSL melons offer growers increased harvest flexibility, although often at the expense of culinary quality. The flesh of LSL melons tends to be firmer and chewier, leading to reduced consumption. This decreased demand is also attributable to several contamination and food safety events. Future breeding efforts must strike a balance between the traits prioritized by growers and those desired by consumers.

1.2 CLIMATE CHANGE

Climate change is considered to be one of the most significant threats to global food production. Extreme weather conditions such as high temperatures, heavy rainfall, droughts, and cold snaps can disrupt crop growth and reduce yields. Climate change refers to long-term alterations in the statistical properties of climatic factors that persist over extended periods (Kopeć, 2024). While the root causes of climate change are complex, leading theories point to the documented rise in global temperature (global warming) and anthropogenic influences on atmospheric composition and land use (IPCC, 2022). In addition to rising average temperatures, climate change is expected to lead to increased rainfall and more unpredictable extreme weather events, such as flooding, droughts, and cold snaps. Climate change can significantly impact agricultural suitability and alter the distribution of production zones. For example, rising temperatures may push crop cultivation into regions in the northern hemisphere (near the Arctic), which were previously limited by cold temperatures (Zsögön et al., 2022).

To address the challenges imposed by climate change, breeders must develop cultivars that not only meet local growing needs but also exhibit stability in diverse environmental conditions. The strong influence of GE interactions on desirable traits, which are typically inherited quantitatively, remains a significant impediment for melon breeders and geneticists (Ajay et al., 2020; Xu et al., 2017; Zhang et al., 2020). Traits evaluated under one environmental condition often do not replicate under different conditions. To identify and select alleles that remain stable across diverse environments, breeders must integrate advanced technologies, including genomic prediction, big data, deep learning, advanced sequencing, and high-throughput phenotyping (Eltaher et al., 2021; Tian et al., 2021).

1.3 DROUGHT

The negative impacts of drought on crop productivity and quality are well documented (Minhas et al., 2017; Meza et al., 2020). Drought disproportionately impacts different melon production regions. For instance, in Rajasthan, India, where melons are considered naturalized, cultivation is heavily dependent on monsoon rains. In 1992, the Indo-U.S. *Cucumis* Expedition documented the complete loss of cucumber landraces due to repeated droughts (McCreight et al., 2020).

Drought inhibits plant growth by impeding water uptake, disrupting nitrogen metabolism, and inducing oxidative stress. As a result, photosynthesis is compromised due to stomatal closure, reduced transpiration, and a lowered concentration of carbon dioxide (Ansari et al., 2019). Additionally, drought stress causes cellular damage in plants, triggering the activation of stress-response molecules, including heat shock proteins, proteinase inhibitors, thiol proteases, and osmotin.

Developing drought-resilient melon varieties requires breeders to screen extensive germplasm collections. In a previous study, Rehman et al. (2024) employed cluster analysis to classify numerous melon genotypes into five groups based on their drought resistance and sensitivity. Their findings identified mel-46, mel-58, and mel-15 as drought-resistant genotypes, which may be further evaluated for use in breeding programs. The diverse responses of melon accessions to drought stress are likely governed by multiple genes, highlighting the need for further investigation.

1.4 SALINITY

High water quality is crucial for the successful cultivation of all crops, including melons. As climate change intensifies water scarcity, growers may resort to using saline water for irrigation, which negatively impacts crops at electrical conductivity levels above 2.2 dS m^{-1} (Freitas et al., 2014; Kim et al., 2016; Medeiros et al., 2012). Salt stress can lead to the accumulation of sodium ions, disruption of cellular balance, and dehydration in plants. It is known to impair seed germination, growth, and fruit development. Additionally, salt exposure can increase oxidative stress in melons, potentially damaging cellular and metabolic functions. Limited water availability and increased salinity are projected to pose long-term problems in the melon production areas of Arizona and California (McCreight et al., 2020).

Considering the economic importance of melon, breeders must develop cultivars with increased salinity tolerance. Several melon genotypes exhibit salinity tolerance (Kusvuran, 2012) for instance, da Silva et al. (2021) demonstrated that certain melon accessions were more productive than commercial cultivars when grown under salt stress conditions. Moreover, a comparison between salt-sensitive and salt-tolerant melon cultivars revealed that genotypes with increased histidine content and/or the ability to accumulate sodium (likely in the vacuoles) display enhanced tolerance to salt stress (Chevilly et al., 2021).

1.5 COLD STRESS

Early in the season, melon plants are susceptible to chilling and frost injury. This damage primarily occurs during transitional periods, such as spring and fall. Many growers circumvent this damage by implementing tunnels at the seedling stage. Low temperatures adversely affect plant and fruit development by altering cell membrane structure and fluidity, leading to the excessive accumulation of reactive oxygen species, reduced water availability, impaired photosynthetic efficiency, and slowed biochemical reactions (Satyakam et al., 2022; Diao et al., 2024).

A genetic diversity panel identified two primary groups of *C. melo*, *melo* and *agrestis*, among which *agrestis* accessions exhibited the strongest tolerance to chilling stress (Hou et al., 2018). Recently, Li et al. (2023) found that cultivars with high chilling resistance tend to have lower metabolite contents than their sensitive counterparts. The metabolism of glutathione, arginine, and proline in melon leaves primarily regulates responses to chilling stress. Transcription factors, such as MYB, ERF, MADS-box, and bZIP, play a crucial role in enhancing the cold tolerance of melon (Diao et al., 2024). In the face of increasingly unpredictable environmental conditions, understanding the genetic basis of chilling tolerance is vital for future melon breeding.

1.6 FUNGAL PATHOGENS

Melon cultivation faces significant threats from several fungal pathogens, including Fusarium wilt, powdery mildew, gummy stem blight (GSB), and downy mildew. Fusarium wilt, caused by *F. oxysporum* f. sp. *melonis* (Fom), is one of the most damaging soil-borne diseases of melon, with severe epidemics capable of causing total crop failure (Kalia et al., 2017; Sousaraei et al., 2018). Resistance to the four physiological races, designated 0, 1, 2, and 1.2, is primarily conferred by *Fom-1* and *Fom-2*, which provide resistance to races 0 and 2, and 0 and 1, respectively.

Several foliar diseases significantly impact global melon production. Cucurbit powdery mildew, one of the most devastating fungal diseases, is caused by *Podosphaera xanthii* (the primary causal agent) and *Golovinomyces cichoracearum* (Shahwar et al., 2024). Based on their virulence patterns on known resistance lines, a total of 21 *P. xanthii* races have been identified on melon (Cui et al., 2022; McCreight, 2006). Downy mildew also targets melon leaves, forming distinct, angular, chlorotic lesions on the leaf surface. The causal organism is an obligate oomycete known as *Pseudoperonospora cubensis*. A recent survey of *P. cubensis* populations using *C. melo* sentinel plots across nine U.S. states identified two genotypes capable of causing infection: Clade 1/mating type A2 (88% of isolates) and Clade 2/mating type A1 (12% of isolates) (Toporek et al., 2023). This disease typically occurs in the summer and fall seasons. Disease control efforts heavily rely on fungicides due to the complexity of genetic resistance. Finally, GSB, caused by the ascomycete fungus *Didymella bryoniae* (Auersw.) Rehm, is implicated in immense crop damage and can be transmitted via seed. This pathogen, which has *Phoma cucurbitacearum* (Fr.) Sacc. as its anamorph state, is known to significantly reduce melon yields worldwide (Shahwar et al., 2024). Like downy mildew, control against gummy stem blight relies on chemical applications.

1.7 VIRAL PATHOGENS

Viral diseases in crops are estimated to cause approximately $30 billion in annual losses worldwide (Rabadán et al., 2023). Many virus resistance traits are polygenic in nature, complicating breeding efforts. These traits are strongly influenced by environmental interactions, particularly the interactions between host plants and insect vectors (Argyris et al., 2015). Moreover, viruses can evolve rapidly, allowing them to overcome resistance traits introduced through breeding programs.

Several major vectors are responsible for the distribution of viruses in melons. Aphids are known to transmit CABYV (genus *Polerovirus*), WMV (genus *Potyvirus*), ZYMV (genus *Potyvirus*), papaya ringspot virus (PRSV; genus *Potyvirus*), Moroccan watermelon mosaic virus (MWMV; genus *Potyvirus*), and CMV (genus *Cucumovirus*). Whiteflies vector CYSDV (genus *Crinivirus*), beet pseudoyellows virus (BPYV; genus *Closterovirus*), and cucumber vein yellowing virus (CVYV; genus *Ipomovirus*). Finally, the soil fungus *Olpidium bornovanus* spreads melon necrotic spot virus (MNSV; genus *Carmovirus*) (Rodríguez-Hernández et al., 2012). To improve resistance to viruses, breeders must screen extensive collections of accessions under various environmental conditions.

1.8 INSECTS

A wide variety of insects and nematodes can damage melon plants both directly and through the transmission of viral diseases. Aphids transmit numerous, destructive potyviruses. Leaf miner larvae damage leaves, significantly reducing photosynthetic activity and leading to premature leaf drop. Cucumber beetles damage multiple plant organs by feeding on leaves and scraping the fruit rind, often rendering the fruit unmarketable. These beetles also vector certain bacteria, which cause wilt in melons. In recent decades, sweet potato whitefly (*Bemisia tabaci*) has emerged as a significant global pest. It serves as a vector for several viral diseases, including CYSDV, and can devastate melon crops through feeding damage alone. Damage caused by the fly reduces the host plant's photosynthetic activity, leading to the development of sooty mold (McCreight et al., 2020).

REFERENCES

Ajay, B. C., S. K. Bera, A. L. Singh, Narendra Kumar, K. Gangadhar, and Praveen Kona. 2020. "Evaluation of Genotype × Environment Interaction and Yield Stability Analysis in Peanut under Phosphorus Stress Condition Using Stability Parameters of AMMI Model." *Agricultural Research* 9 (4): 477–86. https://doi.org/10.1007/s40003-020-00458-3.

Ansari, Waquar Akhter, Neelam Atri, Javed Ahmad, Mohammad Irfan Qureshi, Bijendra Singh, Ram Kumar, Vandna Rai, and Sudhakar Pandey. 2019. "Drought Mediated Physiological and Molecular Changes in Muskmelon (*Cucumis melo* L.)." *PLoS One* 14 (9): e0222647. https://doi.org/10.1371/journal.pone.0222647.

Argyris, Jason M., Marta Pujol, Ana Montserrat Martín-Hernández, and Jordi Garcia-Mas. 2015. "Combined Use of Genetic and Genomics Resources to Understand Virus Resistance and Fruit Quality Traits in Melon." *Physiologia Plantarum* 155 (1): 4–11. https://doi.org/10.1111/ppl.12323.

Burger, Yosef, and Arthur A. Schaffer. 2007. "The Contribution of Sucrose Metabolism Enzymes to Sucrose Accumulation in *Cucumis melo*." *Journal of the American Society for Horticultural Science* 132 (5): 704–12. https://doi.org/10.21273/JASHS.132.5.704.

Chang, Chi-I., Chang-Hung Chou, Ming-Huei Liao, Tz. Min Chen, Chia-Hsin Cheng, Rista Anggriani, Chung-Pao Tsai, Hsin-I. Tseng, and Hsueh-Ling Cheng. 2015. "Bitter Melon Triterpenes Work as Insulin Sensitizers and Insulin Substitutes in Insulin-Resistant Cells." *Journal of Functional Foods* 13: 214–24. https://doi.org/10.1016/j.jff.2014.12.050.

Chevilly, Sergio, Laura Dolz-Edo, Gema Martínez-Sánchez, Luna Morcillo, Alberto Vilagrosa, José M. López-Nicolás, José Blanca, Lynne Yenush, and José M. Mulet. 2021. "Distinctive Traits for Drought and Salt Stress Tolerance in Melon (*Cucumis melo* L.)." *Frontiers in Plant Science* 12: 777060. https://doi.org/10.3389/fpls.2021.777060.

Cui, Lei, Lampros Siskos, Chen Wang, Henk J. Schouten, Richard G. F. Visser, and Yuling Bai. 2022. "Breeding Melon (*Cucumis melo*) with Resistance to Powdery Mildew and Downy Mildew." *Horticultural Plant Journal* 8 (5): 545–61. https://doi.org/10.1016/j.hpj.2022.07.006.

da Silva, Fernando Henrique Alves, Patrícia Lígia Dantas de Morais, Nildo da Silva Dias, Glauber Henrique de Sousa Nunes, Marciana Bizerra de Morais, Marlenildo Ferreira Melo, and Maria Tereza de Albuquerque Nascimento. 2021. "Physiological Aspects of Melon (*Cucumis melo* L.) as a Function of Salinity." *Journal of Plant Growth Regulation* 40 (3): 1298–314. https://doi.org/10.1007/s00344-020-10190-5.

Diao, Qiannan, Shoubo Tian, Yanyan Cao, Dongwei Yao, Hongwei Fan, Xuejun Jiang, Wenxian Zhang, and Yongping Zhang. 2024. "Physiological, Transcriptomic, and Metabolomic Analyses of the Chilling Stress Response in Two Melon (*Cucumis melo* L.) Genotypes." *BMC Plant Biology* 24 (1): 1074. https://doi.org/10.1186/s12870-024-05773-3.

Dogimont, C., and N. Sari. 2022. "2022 Gene List for Melon." *Cucurbit Genetics Cooperative Report* 45: 1–47.

Eltaher, Shamseldeen, P. Stephen Baenziger, Vikas Belamkar, Hamdy A. Emara, Ahmed A. Nower, Khaled F. M. Salem, Ahmad M. Alqudah, and Ahmed Sallam. 2021. "GWAS Revealed Effect of Genotype × Environment Interactions for Grain Yield of Nebraska Winter Wheat." *BMC Genomics* 22 (1): 2. https://doi.org/10.1186/s12864-020-07308-0.

FAOSTAT [Database]. 2025. *Production*. Accessed May 9, 2025. https://www.fao.org/faostat/en/#data/QCL/visualize.

Freitas, Ludimilla D. A., Vladimir B. Figueirêdo, Francisco de Q. Porto Filho, Porto Filho, Porto Filho, J. C. Costa, and E. M. Cunha. 2014. "Crescimento e Produção do Meloeiro Cultivado Sob Diferentes Níveis de Salinidade e Nitrogênio." *Revista Brasileira de Engenharia Agrícola e Ambiental* 18 (Suppl.): 20–6. https://doi.org/10.1590/1807-1929/agriambi.v18nsupps20-s26.

Garcia-Mas, Jordi, Andrej Benjak, Walter Sanseverino, Michael Bourgeois, Gisela Mir, Víctor M. González, Elizabeth Hénaff, Francisco Câmara, Luca Cozzuto, Ernesto Lowy, Tyler Alioto, Salvador Capella-Gutiérrez, Jose Blanca, Joaquín Cañizares, Pello Ziarsolo, Daniel Gonzalez-Ibeas, Luis Rodríguez-Moreno, Marcus Droege, Lei Du, Miguel Alvarez-Tejado, Belen Lorente-Galdos, Marta Melé, Luming Yang, Yiqun Weng, Arcadi Navarro, Tomas Marques-Bonet, Miguel A. Aranda, Fernando Nuez, Belén Picó, Toni Gabaldón, Guglielmo Roma, Roderic Guigó, Josep M. Casacuberta, Pere Arús, and Pere Puigdomènech. 2012. "The Genome of Melon (*Cucumis melo* L.)." *Proceedings of the National Academy of Sciences of the United States of America* 109 (29): 11872–7. https://doi.org/10.1073/pnas.1205415109.

Gómez-García, Ricardo, Débora A. Campos, Cristóbal N. Aguilar, Ana R. Madureira, and Manuela Pintado. 2020. "Valorization of Melon Fruit (*Cucumis melo* L.) By-Products: Phytochemical and Biofunctional Properties with Emphasis on Recent Trends and Advances." *Trends in Food Science & Technology* 99: 507–19. https://doi.org/10.1016/j.tifs.2020.03.033.

Hou, Juan, Ya-Feng Zhou, Lu-Yin Gao, Yan-Ling Wang, Lu-Ming Yang, Hua-Yu Zhu, Ji-Ming Wang, Sheng-Jie Zhao, Chang-Sheng Ma, Shou-Ru Sun, and Jian-Bin Hu. 2018. "Dissecting the Genetic Architecture of Melon Chilling Tolerance at the Seedling Stage by Association Mapping and Identification of the Elite Alleles." *Frontiers in Plant Science* 9: 1577. https://doi.org/10.3389/fpls.2018.01577.

IPCC. 2022. "Impacts, Adaptation and Vulnerability. Contribution of Working Group II to the Sixth Assessment Report of the Intergovernmental Panel on Climate Change." *Climate*

Change, edited by H.-O. Portner, D. C. Roberts, M. Tignor, E. Poloczanska, K. Mintenbeck, A. Alegría, M. Craig, S. Langsdorf, S. Loschke, V. Moller et al. Cambridge, UK; New York, NY: Cambridge University Press.

Kalia, Anu, Sat Pal Sharma, and Vinod Kumar Vashisht. 2017. "Scanning Electron Microscopy Study of Root Tissue of Muskmelon: Transferring Fusarium Wilt Resistance from Snapmelon to Muskmelon." *Journal of Applied & Natural Science* 9 (3): 1317–23. https://doi.org/10.31018/jans.v9i3.1360.

Kim, Hakkwan, Hanseok Jeong, Jihye Jeon, and Seungjong Bae. 2016. "Effects of Irrigation with Saline Water on Crop Growth and Yield in Greenhouse Cultivation." *Water* 8 (4): 2–9. https://doi.org/10.3390/w8040127.

Kopeć, Przemysław. 2024. "Climate Change-The Rise of Climate-Resilient Crops." *Plants* 13 (4): 490. https://doi.org/10.3390/plants13040490.

Kusvuran, S. 2012. "Effects of Drought and Salt Stresses on Growth, Stomatal Conductance, Leaf Water and Osmotic Potentials of Melon Genotypes (*Cucumis melo* L.)." *African Journal of Agricultural Research* 7 (5): 775–81. https://doi.org/10.5897/AJAR1.

Li, Lili, Qiong Li, Bin Chen, Jiyu Wang, Fei Ding, Panqiao Wang, Xiuyue Zhang, Juan Hou, Renren Luo, Xiang Li, Jingwen Zheng, Sen Yang, Luming Yang, Lei Zhu, Shouru Sun, Changsheng Ma, Qin Li, Ying Li, and Jianbin Hu. 2023. "Identification of Candidate Genes That Regulate the Trade-Off between Seedling Cold Tolerance and Fruit Quality in Melon (*Cucumis melo* L.)." *Horticulture Research* 10 (7): uhad093. https://doi.org/10.1093/hr/uhad093.

Manchali, Shivapriya, Kotamballi N. Chidambara Murthy, C. Vishnuvardana, and Bhimanagouda S. Patil. 2021. "Nutritional Composition and Health Benefits of Various Botanical Types of Melon (*Cucumis melo* L)." *Plants* 10 (9): 1755. https://doi.org/10.3390/plants10091755.

McCreight, J., Angela Davis, and Kathleen Reitsma. 2020. *Melon Crop Vulnerability Statement 2020*. Accessed July 9, 2025. https://www.ars-grin.gov/documents/cgc/cvs/2020-Melon%20Crop%20Vulnerability%20Statement.pdf.

McCreight, James D. 2006. "Melon-Powdery Mildew Interactions Reveal Variation in Melon Cultigens and *Podosphaera xanthii* Races 1 and 2." *Journal of the American Society for Horticultural Science* 131 (1): 59–65. https://doi.org/10.21273/JASHS.131.1.59.

Medeiros, Damiana C. de, José F. de Medeiros, Murillo A. G. Barbosa, Roberto C. F. Queiroga, Francisco de A. Oliveira, and Wallace E. de S. Freitas. 2012. "Crescimento do Melao Pele de Sapo, em Niveis de Salinidade e Estagio de Desenvolvimento da Planta." *Revista Brasileira de Engenharia Agrícola e Ambiental* 16 (6): 647–54. https://doi.org/10.1590/S1415-43662012000600009.

Meza, Isabel, Stefan Siebert, Petra Döll, Jürgen Kusche, Claudia Herbert, Ehsan Eyshi Eyshi Eyshi Rezaei, Hamideh Nouri, Helena Gerdener, Eklavyya Popat, Janna Frischen, Gustavo Naumann, Jürgen V. Vogt, Yvonne Walz, Zita Sebesvari, and Michael Hagenlocher. 2020. "Global-Scale Drought Risk Assessment for Agricultural Systems." *Natural Hazards & Earth System Sciences* 20 (2): 695–712. https://doi.org/10.5194/nhess-20-695-2020.

Minhas, Paramjit S., Jagadish Rane, and Ratna K. Pasala. 2017. "Abiotic Stresses in Agriculture: An Overview." In *Abiotic Stress Management for Resilient Agriculture*, edited by Paramjit S. Minhas, Jagadish Rane, and Ratna K. Pasala. Singapore: Springer, 3–8. https://doi.org/10.1007/978-981-10-5744-1_1.

Oren, Elad, Asaf Dafna, Galil Tzuri, Ilan Halperin, Tal Isaacson, Meital Elkabetz, Ayala Meir, Uzi Saar, Shachar Ohali, Thuy La, Cinta Romay, Yaakov Tadmor, Arthur A. Schaffer, Edward S. Buckler, Roni Cohen, Joseph Burger, and Amit Gur. 2022. "Pan-genome and Multi-parental Framework for High-Resolution Trait Dissection in Melon (*Cucumis melo*)." *Plant Journal: For Cell & Molecular Biology* 112 (6): 1525–42. https://doi.org/10.1111/tpj.16021.

Rabadán, M. P., M. Juárez, and P. Gómez. 2023. "Long-Term Monitoring of Aphid-Transmitted Viruses in Melon and Zucchini Crops: Genetic Diversity and Population Structure of Cucurbit Aphid-Borne Yellows Virus and Watermelon Mosaic Virus." *Phytopathology* 113 (9): 1761–72. https://doi.org/10.1094/PHYTO-10-22-0394-V.

Rehman, Asad, Muhammad Khalid, Jinyang Weng, Pengli Li, S. Rahman, Iftikhar Hussain Shah, Shazma Gulzar, Song Tu, Feng Ningxiao, Qingliang Niu, and Liying Chang. 2024. "Exploring Drought Tolerance in Melon Germplasm through Physiochemical and Photosynthetic Traits." *Plant Growth Regulation* 102 (3): 603–18. https://doi.org/10.1007/s10725-023-01080-3.

Rodríguez-Hernández, Ana M., Blanca Gosalvez, Raquel N. Sempere, Lorenzo Burgos, Miguel A. Aranda, and Verónica Truniger. 2012. "Melon RNA interference (RNAi) Lines Silenced for Cm-eIF4E Show Broad Virus Resistance." *Molecular Plant Pathology* 13 (7): 755–63. https://doi.org/10.1111/j.1364-3703.2012.00785.x.

Satyakam, Z. G., Gaurav Zinta, Rajesh Kumar Singh, and Rajiv Kumar. 2022. "Cold Adaptation Strategies in Plants-An Emerging Role of Epigenetics and Antifreeze Proteins to Engineer Cold Resilient Plants." *Frontiers in Genetics* 13: 909007. https://doi.org/10.3389/fgene.2022.909007.

Schaffer, Arthura A., Beny Aloni, and Edna Fogelman. 1987. "Sucrose Metabolism and Accumulation in Developing Fruit of *Cucumis*." *Phytochemistry* 26 (7): 1883–7. https://doi.org/10.1016/S0031-9422(00)81721-5.

Shahwar, Durre, Zeba Khan, and Younghoon Park. 2024. "Molecular Markers for Marker-Assisted Breeding for Biotic and Abiotic Stress in Melon (*Cucumis melo* L.): A Review." *International Journal of Molecular Sciences* 25 (12): 6307. https://doi.org/10.3390/ijms25126307.

Sharma, Nity, Jaideep Kaur Deol, Gurleen Kaur, Agampreet Kaur, Sat Pal Sharma, and Navraj Kaur Sarao. 2025. "Exploring the Role of Advanced Genomics in Melon Breeding: A Review." *Plant Molecular Biology Reporter*. https://doi.org/10.1007/s11105-025-01552-y.

Sousaraei, N., H. Ramshini, M. Lotfi, and A. Sharzei. 2018. "Marker Assisted Backcrossing for Introgression of *Fusarium Wilt* Resistance Gene into Melon." *Euphytica* 214: 1–10.

Sun, Yang, Jing Wang, Yan Li, Bin Jiang, X. Wang, Wen-Hui Xu, Yu-Qing Wang, Pei-Tao Zhang, Yong-Jun Zhang, and Xiang-Dong Kong. 2022. "Pan-genome Analysis Reveals the Abundant Gene Presence/Absence Variations among Different Varieties of Melon and Their Influence on Traits." *Frontiers in Plant Science* 13: 835496. https://doi.org/10.3389/fpls.2022.835496.

Tian, Shoubo, Qiannan Diao, Yanyan Cao, Dongwei Yao, Wenxian Zhang, Hui Zhang, Xuan Du, and Yongping Zhang. 2024 Dec. 12. "Overview of Research on Virus-Resistant Breeding of Melon." *Frontiers in Plant Science* 15: 1500246. https://doi.org/10.3389/fpls.2024.1500246.

Tian, Zhixi, Jia-Wei Wang, Jiayang Li, and Bin Han. 2021. "Designing Future Crops: Challenges and Strategies for Sustainable Agriculture." *Plant Journal: For Cell & Molecular Biology* 105 (5): 1165–78. https://doi.org/10.1111/tpj.15107.

Toporek, Sean M., Sandra E. Branham, Anthony P. Keinath, and W. Patrick Wechter. 2023. "QTL Mapping of Resistance to Pseudoperonospora cubensis Clade 2, Mating Type A1, in Cucumis melo and Dual-Clade Marker Development." *Theoretical & Applied Genetics* 136 (4): 91. https://doi.org/10.1007/s00122-023-04333-x.

Xu, Y., C. Xu, and S. Xu. 2017. "Prediction and Association Mapping of Agronomic Traits in Maize Using Multiple Omic Data." *Heredity* 119 (3): 174–84. https://doi.org/10.1038/hdy.2017.27.

Zeven, A. C., and J. M. De Wet. 1982. *Dictionary of Cultivated Plants and Their Regions of Diversity: Excluding Most Ornamentals, Forest Trees and Lower Plants*. Wageningen, the Netherlands: PUDOC, 21–196.

Zhang, H., X. Li, H. Yu, Y. Zhang, M. Li, H. Wang, D. Wang, H. Wang, Q. Fu, M. Liu, C. Ji, L. Ma, J. Tang, S. Li, J. Miao, H. Zheng, and H. Yi. 2019. "High-Quality Melon Genome Assembly Provides Insights into Genetic Basis of Fruit Trait Improvement. *iScience* 2 (049): 16–27.

Zhang, Meiping, Yanru Cui, Yun-Hua Liu, Wenwei Xu, Sing-Hoi Sze, Seth C. Murray, Shizhong Xu, and Hong-Bin Zhang. 2020. "Accurate Prediction of Maize Grain Yield Using Its Contributing Genes for Gene-Based Breeding." *Genomics* 112 (1): 225–36. https://doi.org/10.1016/j.ygeno.2019.02.001.

Zsögön, Agustin, Lázaro E. P. Peres, Yingjie Xiao, Jianbing Yan, and Alisdair R. Fernie. 2022. "Enhancing Crop Diversity for Food Security in the Face of Climate Uncertainty." *Plant Journal: For Cell & Molecular Biology* 109 (2): 402–14. https://doi.org/10.1111/tpj.15626.

2 Genetic Diversity and Classification of Melon (*Cucumis melo* L.)

Nebahat Sari, Ilknur Solmaz, and Catherine Dogimont

2.1 INTRODUCTION

Melon (*Cucumis melo* L.), which belongs to the *Cucurbitaceae* family, is a diploid ($2n = 2x = 24$), cross-pollinated plant, with a relatively small genome (454 Mb; Garcia-Mas et al., 2012), high intraspecific genetic variation, and morphological diversity (Pitrat, 2017). It is a vegetable crop of high economic value worldwide, cultivated mainly in tropical and subtropical climates for its sweet, refreshing taste and high nutritional value (Decker-Walters et al., 2002; Garcia-Mas et al., 2004; Pech et al., 2008; Frary et al., 2013; Grumet et al., 2021; Lija and Beevy, 2021; Solmaz et al., 2021; Shahwar et al., 2023; Naznin et al., 2024). A total of 28.5 million tons of melons are produced on 1 million hectares worldwide. On a continental basis, the highest melon production is in Asia (21.9 million tons), followed by the Americas (3.5 million tons), Europe (1.7 million tons), Africa (1.2 million tons), and Oceania (0.22 million tons). China is the leading producer (14.2 million tons), followed by Türkiye (1.6 million tons) and India (1.5 million tons; Figure 2.1; FAO, 2022, http://www.fao.org/faostat/en/#data/QC, accessed 21 November 2024).

Melons are very beneficial for human health due to their richness in ascorbic acid, carotene, folic acid, potassium, and various other bioactive compounds. β-carotene (provitamin A) is the primary carotenoid in orange-fleshed melons and serves as a vital antioxidant that strengthens the immune system. In addition, it has been reported to reduce the risk of cancer, prevent cardiovascular disorders, cataract, and night blindness (Lester and Shellie, 2004; Hodges and Lester, 2011; Rolim et al., 2020; Kaushik, 2023).

Plant gene banks, which conserve plant genetic resources *in situ* and *ex situ* conditions, are responsible for the collection, conservation, characterization, documentation, and distribution of plant genetic resources (Anglin et al., 2018). Melon is usually preserved in *ex situ* gene banks, and there are around 10,000 melon accessions in national gene banks in various countries (Grumet et al., 2021). The largest melon seed collections in the world are in Russia, the United States, France, Spain, China, Brazil, Türkiye, Germany, Taiwan, and Bulgaria, among others. A gene bank established between Norway and the north pole (Svalbard Global Seed Vault) also

DOI: 10.1201/9781003715757-2

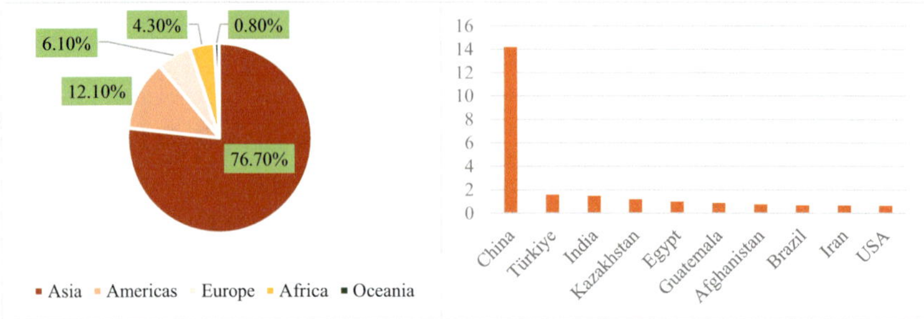

FIGURE 2.1 Melon production amounts (tons) by continents (left) and countries (right; FAOSTAT, 2022, accessed 21 November 2024).

preserves seeds under natural conditions as part of a contract against the loss of genetic resources due to natural (flood, earthquake, fire, erosion, etc.) or man-made disasters such as wars.

Biodiversity studies on melon were initially started with morphological characterizations; then, studies were continued with genetic and biochemical characterizations with molecular marker technologies, such as randomly amplified polymorphic DNA (RAPD), restriction fragment length polymorphism (RFLP), amplified fragment length polymorphism (AFLP), inter-simple sequence repeat (ISSR), simple sequence repeat (SSR), and single nucleotide polymorphism (SNP). The last subsection of this chapter lists studies conducted on melon morphologic and genetic characterization in the world to date.

2.2 CLASSIFICATION OF MELON, BOTANICAL, AND HORTICULTURAL GROUPS

The Cucurbitaceae family is among the most economically important plant groups. It belongs to the order Cucurbitales, containing 96 genera and approximately 1000 species. This botanical family includes many widely cultivated crops such as watermelon (*Citrullus lanatus* [Thunb.] Matsum. & Nakai), melon (*Cucumis melo* L.), cucumber (*Cucumis sativus* L.), gourds, pumpkins, squashes, and zucchinis (*Cucurbita* spp.), and species such as *Lagenaria ssp.*, *Benincasa* ssp., *Luffa* ssp., *Sicyos* ssp., and *Cyclanthera* ssp., among others, which are used as food and ornamental plants in various parts of the world (Chomicki et al., 2019).

Melon has a long and confusing taxonomic history, such as many other crop. The name "*Cucumis melo*" was introduced by Linnaeus (1753), who described a large number of plant species. Melon, an important member of the Cucurbitaceae family, displays high polymorphism in terms of botanical varieties and fruit types (Nunez-Palenius et al., 2008), and according to the shape of the pubescence on the ovary, the long and soft-tempered ones were classified as *C. melo* subsp. *melo*; the short and more pronounced ones were classified as *C. melo* subsp. *agrestis* (Naudin) Pangalo (Kirkbride, 1993). Munger and Robinson (1991) reported that there are 7 subspecies of *C. melo*. Of these subspecies, *C. melo* var. *agrestis* has a weak stem and monoecious structure, is sought after in Africa and Asia, and grows as a wild. *C. melo* var. *cantalupensis* has medium or large sweet, juicy, and aromatic fruits. *C. melo* var. *inodorus* is a winter melon; these melons with large and odorless fruits have a very long shelf life. *C. melo* var. *flexuosus* has very long fruits and is consumed like cucumber. *C. melo* var. *conomon* has smooth, white, and thin-skinned sweet fruits; its fruit is consumed with its peel. *C. melo* var. *chito* and *dudaim* have small fruits; they are used in pickles and as ornamental plants. *C. melo* var. *momordica* has a monoecious flower structure, and its ovary is widespread and wide; the fruits are tasteless and thin-skinned.

Pitrat (2017) classified melon into 19 groups according to their geographical origin, morphology, and horticultural traits. Following this classification, the main characteristics of the 19 melon groups (Pitrat, 2017; Solmaz et al., 2021) cultivated and collected from their wild natural habitats will be mentioned. Two groups were added to previous horticultural classifications; *indicus* and *kachri*, the *inodorus* group, was divided into *ibericus*, *casaba*, and *inodorus*, and the previous group, *reticulatus*, was merged with *cantalupensis*. The most intensively cultivated melon groups in the world are the *cantalupensis*, *inodorus*, *casaba*, and *ibericus*:

Group *acidulus:* The fruits are sugarless and aroma-less and mainly cultivated in India and Sri Lanka.

Group *agrestis:* It is a wild melon with very small fruits, found in various parts of the world, in Asia, Africa, and Australia. Because its plant is very strong and it sets a lot of fruit, the entire plant can be used to protect cultivated melon fruits from the sun during roadside sell in some countries such as Türkiye.

Group *ameri:* This group is mainly cultivated in Asia from Türkiye to western China with their excellent fruit quality such as crisp flesh, texture, and sweet and juicy flesh characteristics.

Group *cantalupensis:* This group was first developed in Europe likely with melons collected from Van Province in Türkiye (Pangalo, 1951). Later, diversification and selection in the United States led to American cantaloupe melons. Sweet, juicy, aromatic, early, odorous, medium-sized yellow seeds without gelatinous coating, short- to medium-shelf-life summer melons are in this group. Galia, Charentais, Hale's Best, and many commercial Galia hybrids grown in greenhouses for the early season are important representatives of this group.

Group *cassaba:* It is a group that includes many cultivated varieties such as Kırkağac, Yuva, Hasanbey, and Kuscular Turkish melons. This group has more or less wrinkled, no ribs, no veins, no netting, a non-dehiscent peduncle, thick skin, thick light-green flesh, high sugar content, late maturation, and long shelf life.

Group *chandalak:* It is a group cultivated from Central Asia to India that usually has a large pistil scar, thin exocarp, usually green (but sometimes orange or white) flesh, medium sugar content, low aroma, large seeds in a gelatinous sheath, early maturation, and short shelf life.

Group *chate:* It has sugar-free and flavorless fruits.

Group *chinensis:* It has pear-shaped fruits. The sugar content is moderate, and it has aroma-less or slightly aromatic fruits.

Group *chito:* Its fruits are small, sugarless, and aroma-less. It is used in pickling or as an ornamental plant.

Group *conomon:* The Conomon group is grown in Far Eastern countries, such as China and Japan. The fruits are harvested when immature and consumed raw like cucumbers or are used in pickling. Tokyo Wase Shiro Uri and Ko Shiro Uri are representatives of this group. It has a smooth and thin-skinned fruit. The white and hard fruit flesh is sugarless and aroma-less. It does not show climacteric properties.

Group *dudaim:* "Fragrant Melon" or "Queen Anne's Pocket Melon". In the years before perfume was invented, it was carried in pockets and kept in rooms and wardrobes for a few months to spread a beautiful scent because it had a long shelf life. It has small round/ovate fruit that can fit in the palm of the hand and are sugar-free, however flavorful, fruit. It is also grown as an ornamental plant. Türkiye is an important center of origin.

Group *flexuosus:* It is a melon group with grooved, sugarless, and aroma-less fruits, also called "snake melon" or "snake cucumber", and is used fresh or in pickles, like cucumber, because it is a sugar-free melon grown in some Asian, African, and European countries.

Group *ibericus:* Mainly popular in Spain, as well as throughout the Mediterranean region and North and South America. This group of late-season and long-shelf melons has large fruits, thick rinds, thick light green (sometimes light orange) juicy flesh, high sugar content, and large yellow seeds. Piel de Sapo, Amarillo, and Tendral are representatives of this group.

Group *inodorus:* Includes melons with low aroma and long shelf life that do not belong to group cassaba or group ibericus. The fruit skin is not wrinkled, there are no ribs, no veins, with or without netting, non-dehiscent peduncles, a thick rind, thick light green flesh, high sugar content, late maturing, and long-shelf-life melons. Honeydew and Earl's are the most important representatives of this group.

Group *indicus:* A group of plants grown in India with large pistil scars; orange (rarely green) flesh; sweet (high °Brix); thick, juicy, firm flesh; and large seeds with gelatinous coatings.

Group *kachri:* According to Pitrat (2017), this group forms an intermediate group between wild melons and locally cultivated species belonging to different groups. It is quite heterogeneous and can be best described as "major agrestis". It is cultivated in Rajasthan, a semi-desert region of India. It is considered to be drought tolerant.

Group *makuwa:* Its white fleshy fruits are aroma-less but sweet and tasty and show climacteric properties. South Korea is an important cultivation country of Makuwa group. It has smooth, thin, white, yellow, or light green fruit skin.

Group *momordica:* The fruits are tasteless and thin-skinned. The sugar content is low, and the aroma is low and mainly cultivated in India and southeast Asia.

Group *tibish:* Its fruits are small and hard; the flesh is white, sugar-free, and flavorless. It is harvested before ripening and consumed raw like cucumber and is pickled or added to cooked foods. It is grown in East Africa.

2.3 ORIGIN CENTER AND DOMESTICATION

Melon's origin and history have generated much interest among scientists for more than a century. The era of genomics has given rise to some surprising discoveries. However, many uncertainties remain about the domestication process, and we can expect further unexpected developments. Our aim is not to retrace the whole story, which has been described previously (Stepansky et al., 1999; Janick, 2007; Burger et al., 2010; Endl et al., 2018), but to integrate the new data into our understanding models.

2.3.1 The Question of the Melon's Center of Origin

Three continents have been successively proposed as the center of origin of melon: Africa, Asia, and Australia/Oceania/New Guinea. For many years, Africa was considered the center of origin of melon, essentially because most wild species in the genus *Cucumis* were native to Africa, mainly from southern and eastern Africa (Kirkbride, 1993). Asian countries, including Türkiye, Iran, India, Afghanistan, and China, were considered the secondary gene center (Solmaz et al., 2021).

In 2010, a comprehensive phylogenetic analysis based on combined nuclear and plastid sequence data highly remodeled the genus *Cucumis* (Sebastian et al., 2010; Telford et al., 2011). To the wild African *Cucumis* species were added more than 20 new species, especially from Asian, Malaysian, and Australian origins. The high genomic similarity between melon (*C. melo* $2n = 24$) and cucumber (*Cucumis sativus* $2n = 14$), even if they do not share the same chromosome number, has suggested an Asian origin for melon. In contrast to melon, the cucumber progenitor, *C. sativus* var. *hardwickii*, has been clearly identified and originates from the Himalayas (Qi et al., 2013). Finally, a major result in recent years was to identify *Cucumis picrocarpus* and *Cucumis trigonus* Roxb., as the closest relatives of *C. melo.* A phylogenetic analysis has dated the divergence between *Cucumis melo* and its closest wild relatives and potential progenitors to 2.8 Ma (Endl et al., 2018). *C. picrocarpus* is an annual species, native to Australia (Telford et al., 2011). Interestingly, Telford et al. (2011) described that wild *C. melo* and *C. picrocarpus* are widespread and still coexist in Australia. *C. trigonus* is a perennial Indian species, which was described by Roxburgh (1832; in Endl et al., 2018). The accession named *Cucumis trigonus* Ames 24297 has been used in a few publications (Sebastian et al., 2010; Díaz et al., 2017), but the authors now consider that it was misidentified and does not belong to *C. trigonus* Roxb. but is instead a wild *C. melo* (Endl et al., 2018). It should be noted that this phylogenetic study (Endl et al., 2018) was largely based on herbarium leaf samples assessed by botanists and that very few, if any, seeds are available in gene banks, even though this material would greatly benefit from collection and study.

Another major result from the molecular dating analysis was that *C. melo* split into two lineages around 1.9 million years ago: *C. melo* subsp. *Melo*, the Asian/Australian subspecies, and a subspecies *C. melo* subsp. *meloides*, restricted to Africa (Endl et al., 2018). The *C. melo* subsp. *melo* comprises wild accessions from India, Pakistan, the Philippines, Myanmar, the Maldives, Madagascar, Iran, Iraq, and Israel, as well as four wild Australian melon accessions and one from New Guinea, while wild African melons were included in *C. melo* subsp. *meloides*. The wild African clade split into East, West, and Cape Verde lineages (Endl et al., 2018).

Resequencing data analysis from wild accessions collected in Sudan also showed a split into two or three genetic subgroups, demonstrating the richness of this African gene pool, which needs to be explored further (Wojcik et al., 2024).

2.3.2 Two Independent Domestications of Melon

Considering the molecular data available today, researchers agree that at least two independent domestications of melon events took place, one in Asia and one in East Africa, as already proposed by Naudin (1859). The dating of the *C. melo* divergence between the Asian (*C.melo* subsp. *melo*) and African (*C. melo* subsp. *meloides*) lineages to 1.9 million years ago (Endl et al., 2018), at a time well before plant domestication, which occurs within the last 10,000 years, means that the two domestications are necessarily independent and originated from two distinct wild gene pools that have evolved independently over as many generations.

The Asian lineage has given rise to a very wide diversity of land races, from the Far East, India, South and Central Asia, and the Mediterranean basin, as well as western varieties, while the African lineage has led to cultivars still grown in East Africa, particularly Sudan (Serres-Giardi and Dogimont 2012; Endl et al., 2018; Zhao et al., 2019).

Genomic data highlighted the proximity of wild forms of *C. melo* collected in Africa from east to west to a domesticated form traditionally grown in East Africa, known locally as *tibish* in Sudan (El Tahir and Yousif, 2004) and Bougi in Niger (pers. com. pers. Mustapha Boubacarabdou), showing that the *tibish* type results from a domestication of *C. melo* subsp. *meloides* in Africa (Serres-Giardi and Dogimont, 2012; Chomicki et al., 2019). The presence of numerous private alleles indicates that the Sudanese gene pool has not been widely disseminated (Serres-Giardi and Dogimont, 2012; Agaoua et al., 2022; Rashid et al., 2023), either because this domestication is recent or because the resulting plant material has been lost over time. Interestingly, a recent study showed that watermelon (*Citrullus lanatus* [Thunb.] Matsum. & Nakai), another major cucurbit crop, may have been domesticated in Sudan, where the sweet, white-fleshed forms of the progenitor have been collected, especially the Kordofan region (Renner et al., 2021).

A comprehensive genomic analysis based on the resequencing of 1175 melon accessions suggested that, in addition to the African domestication, two Indian domestication events occurred, derived from distinct wild populations and achieved through diverse genetic pathways (Zhao et al., 2019). This is consistent with the greatest genetic diversity found in Indian and East Asian landraces of *C. melo.*

2.3.3 Dating Melon Domestication

Another major question is the date of domestication, but unlike cereals and other cucurbits, archaeological remains of melon seeds are rare (Fuller et al., 2010). Here, too, recent data continue to rewrite the history of melon.

A recent archaeobotanical analysis of diverse plant food resources unearthed from paddy field excavations in a late Neolithic settlement dated from 5000 to 4300 BP in East China revealed the presence of melon seeds, probably cultivated on a large scale together with rice and a diversity of vegetable remains, among which were *Lagenaria siceraria* var. *macrocarpa* and *Trichosantes kirilowii* (Qiu et al., 2016). In the Indus Valley, the oldest melon remains date back to 4300–3600 BP (Vishnu-Mittre, 1974), but the routes of diffusion through Asia are largely unknown.

Carbonized melon seeds were also discovered in eastern Iran and dated to around 4000 BP (Zohary and Hopf, 2000). In Africa, the oldest findings of melon seeds from Lower Egypt in the Nile valley (Maadi near Cairo) date back to 3700–3500 BP (after cal), showing that melons were a familiar vegetable crop in ancient Egypt, cultivated together with wheat and barley as well as garlic, leek, onion, lettuce, chufa (*Cyperus esculentus* L.), and watermelon (Zohary and Hopf, 2000). Several paintings in Egyptian tombs and wooden models of elongated fruits, dating from approximately 3400 BP, have also been identified as melons (Janick et al., 2007).

Melon cultivation is attested in the archaeobotanical site in southern Spain (Huelva) in 2900 BP, together with different fruit tree species (grapevine, olive, almond, and pomegranate) (Pérez-Jordà et al., 2017). During antiquity, at the beginning of our era, the cultivation of melon has spread in the Mediterranean Basin, as attested by the writings of Pliny (De Re Natura, 1st century CE) as well as by numerous iconographic sources, wall paintings, and mosaics, for example, in Pompei and Ercolano, Italy, and in Tunisia. Two types of melon were attested: one with elongate fruits, unsweet, eaten when immature as vegetables, well established long before the rise of the Roman Empire, and the other, more recent, rounded, ridged, sweet, and aromatic (Paris, 2016).

Large gaps remain to understand how the diffusion of domesticated melons occurred, and in particular, the question remains as to the genetic origin of melons cultivated in the Nile valley as far back as ancient Egypt. They could be the result of an African domestication, as the Kush kingdom (3000–1600 BP) flourished in Nubia along the Nile valley, ensuring strong links between modern-day northern Sudan and Upper Egypt. Much more elongated than the *tibish* type, they could be the result of an earlier or different melon domestication in Africa than that which gave rise to the *tibish* type. Finally, they could also be the result of the Asian domestication, as are modern *flexuosus* grown in Sudan, as are most cultivated or feral melons collected in Africa, South Africa, Zimbabwe, or North Africa (Serres-Giardi and Dogimont, 2012). This last hypothesis would imply extensive exchanges at a very early date. The spread of key food crops from their centers of domestication is thought to have been initiated by the migration of farmers as well as by mobile societies of hunter-gatherers or herder-gatherers (Fuller et al., 2014). Organized plant exploration and the exchange of plants were by-products of trade routes as well as the campaigns of conquerors. The first recorded expedition for new plant resources was that of the famous Queen Hatshepsut, which dates to ancient Egypt in 3500 BP (Janick, 2007). Many investigations are still needed to better understand the melon domestication process.

2.4 DIVERSITY IN WILD, FERAL, AND CULTIVATED MELONS

Various descriptors are used to distinguish the phenotypic diversity in melons. The most commonly used descriptor is the UPOV (International Union for the Protection of New Varieties of Plants) descriptor (UPOV, 2019). Because melon is one of the crops with the highest diversity (especially in fruit characters), the number of characters defining this diversity is also quite high. Excluding disease and pest resistance, a total of 68 characters are defined for melon in the UPOV descriptor, 3 at the seedling level, 8 at the leaf level, 3 at the inflorescence/flowering level, 49 at the immature and mature fruit level, and 5 at the seed level.

The large number of characters used for characterization in melons has necessitated the development of a minimum list of descriptors. Sari and Solmaz (2018) used all UPOV criteria in their study with 193 Turkish melon germplasms, but due to the large number of characters, researchers selected 22 of them as the minimum descriptor list for the Turkish Melon Genetic Resources Atlas, which includes about 400 accessions collected locally. Figures 2.2 and 2.3 show some mature and immature fruit characteristics of melons.

2.5 GENETIC DIVERSITY STUDIES

Melon is one of the most polymorphic species, exhibiting significant variability in plant, leaf, flower, and especially fruit characteristics (Silberstein et al., 1999; Singh

FIGURE 2.2 Melon exhibits extensive diversity in fruit size, shape, grooves, rind dots, and patches, as well as pistil scar (immature and mature fruit examples; pictures provided by N. Sari from INRAE [left] and Cukurova University's [right] germplasms).

FIGURE 2.3 Diversity in the fruit morphology of the wild and cultivated melon germplasms. (The pictures are coming from [1]INRAE and [2]Cukurova University's collections, and provided by [3]N. Sari, 4C. Dogimont.)

et al., 2020; Pandey et al., 2021). Due to the considerable morphological variation in fruit traits, this species is believed to harbor the most diverse varieties within the *Cucumis* genus (Stepansky et al., 1999; Bagheriyan et al., 2015; Dantas et al., 2015; Merheb et al., 2020). As mentioned earlier, several plant taxonomists have proposed intraspecific classifications, particularly based on fruit characteristics (Nunez-Palenius et al., 2008). Within these species, 19 different groups have been identified (Pitrat, 2017). These groups have high levels of diversity in fruit shape, fruit size, fruit and rind color, fruit texture, flavor, aroma, and ripening behavior.

Genetic diversity in melon has been studied using different methods, including phenotypic (Escribano and Lazaro, 2009; Szamosi et al., 2010), isozymic (Akashi et al., 2002; McCreight, 2004), and molecular DNA markers (Katzir et al., 1996; Garcia-Mas et al., 2004). In the late 1990s, morphological and molecular markers were used to determine the genetic diversity in melons in numerous studies, as presented in Table 2.1. Morphological markers have mostly been studied by combining physiological and biochemical properties such as pH, total soluble solids, polysaccharide content, organic acids, and vitamins (Garcia-Mas et al., 2004; Fergany et al., 2011; Escribano et al., 2012; Singh et al., 2020; Chikh-Rouhou et al., 2021). According to literature fruit characteristics (Naudin, 1859; McCollum et al., 1987, 1988; Kirkbride, 1993; Goldman, 2002; Bagheriyan et al., 2015; Dantas et al., 2015; Merheb et al., 2020; Seungbum et al., 2020), fruit maturity stage (Gull, 1988; Miccolis and Saltveit, 1991; Aggelis et al., 1997; Zheng and Wolff, 2000; Fallik et al., 2001; Sakata and Sugiyama, 2002; Shellie and Lester, 2002; Périn et al., 2002; Liu et al., 2004), and flower characteristics (Stepansky et al., 1999; Pandey et al., 2021; Saputra et al., 2024) were investigated within the scope of morphological marker studies.

The different generations of molecular markers (RFLP, RAPD, AFLP, SSR, ISSR) and the novel SNP discovery techniques, based on next-generation sequencing methods, have been used to better understand the infraspecific melon diversity, as detailed in Table 2.1. Several studies focused on a specific group or geographic origin of melons and aimed to better know the genetic diversity available in a local collection of landraces. Thus, the diversity of melons has been actively studied in certain regions, where very large collections have been made (e.g., in Türkiye, Spain, India, and China), whereas information remains scarce for other regions of the world. Some groups, such as the *flexuosus*, have been studied in different parts of the world, in India (Pandey et al., 2021), and all around the Mediterranean basin (Akash et al., 2020; Soltani et al., 2010; Henane et al., 2015; Solmaz et al., 2016; Omari et al., 2018; Ibrahim et al., 2020). Despite this, the evolutionary history of this traditional non-sweet vegetable crops, probably a polyphyletic group, is still poorly understood.

Larger studies have been carried out on a global scale using SSR markers (Serres-Giardi and Dogimont, 2012; Esteras et al., 2013). Worldwide melon collections have been recently resequenced, enabling the identification of a number of SNPs and large-scale genotyping. Population structure analysis showed a subdivision into five to seven major subpopulations, depending on the collection studied (Esteras et al., 2013; Zhao et al., 2019; Chomicki et al., 2019). Wild African and cultivated *tibish* melons were distantly related to most of the other landraces, consistent with the two genomic lineages *C. melo* subsp. *meloides* and *C. melo* subsp. *melo* identified by Endl et al. (2018). In most studies, the further subdivisions of the *C. melo* subsp. *melo* do not necessarily correspond to botanical groups and are more related to geographical origin than belonging to a botanical group. For instance, the distinction between *inodorus* and *cantaloupensis* groups is sometimes ambiguous, despite the significant differences in their morphologies. The high level of admixture in a number of landraces also highlights a complex history involving exchanges, intercrossing, introgression, and extensive breeding work. Interestingly, high genetic diversity was

TABLE 2.1
Morphological and Molecular Markers Used to Determine Genetic Diversity in Melons

Diversity	Group/geographic origin	Marker	References
Molecular	*reticulatus, inodorus, acidulus, conomon; Cucurbita pepo, C. maxima; Citrullus lanatus*	SSR	Katzir et al. (1996)
Molecular	*cantalupensis*, *conomon*, *inodorus*, *flexuosus, momordica, metuliferus*, *C. sativus* var. *sativus*, *C. sativus* var. *hardwickii*, *C. metuliferus*; worldwide	RAPD	Staub et al. (1997)
Molecular	Japanese and American varietal types	RAPD	Garcia et al. (1998)
Molecular	Korea	RAPD	Mo et al. (1998)
Molecular	*conomon*, *chito*, *dudaim*, *agrestis*	RFLP and RAPD	Silberstein et al. (1999)
Molecular and morphological	*cantalupensis*, *inodorus*, *conomon*, *chito*, *dudaim*, *momordica*, *flexuosus*, *agrestis*; worldwide	SSR, RAPD, and seed and flower traits	Stepansky et al. (1999)
Molecular; USA	*cantalupensis*, *inodorus*, *conomon*, and *flexuosus*	AFLP, RAPD, and RFLP	Staub et al. (2000)
Molecular	Spain, Korea, Zimbabwe, and Israel	RAPD, AFLP, and RFLP	Garcia-Mas et al. (2000)
Molecular	*cantalupensis*, *conomon*, *inodorus*, and *flexuosus*; Africa	RAPD	Mliki et al. (2001)
Molecular	*conomon*, *makuwa*; East and South Asia	isozyme	Akashi et al. (2002)
Molecular	*conomon*, *flexuosus*, *inodorus*, and *cantalupensis*; North America	RAPD and SSR	Decker-Walters et al. (2002)
Molecular	*cantalupensis*, *conomon*, *inodorus*, *flexuosus*, and *chito*; Spain	SSR	López-Sesé et al. (2003)
Molecular	*acidulus, adana, reticulatus*, *chate*, and *tibish*; USA, Spain	AFLP, RAPD, and RFLP	Monforte et al. (2003)
Molecular	*cantalupensis and inodorus*; India and China	isozyme	McCreight, 2004
Molecular	*reticulatus, cantalupensis, and inodorus; Citrullus lanatus; Cucurbita maxima, Cucumis sativus;* USA	SSR and RAPD	Ritschel et al., 2004
Molecular and morphological	*cantalupensis*, *inodorus*, and *flexuosus*; Greece	RAPD; fruit traits and powdery mildew resistance	Staub et al. (2004)
Molecular	*cantalupensis*, *conomon*, and *flexuosus*; Japan	RAPD	Nakata et al. (2005)

(Continued)

TABLE 2.1 *(Continued)*
Morphological and Molecular Markers Used to Determine Genetic Diversity in Melons

Diversity	Group/geographic origin	Marker	References
Molecular	*conomon* and *makuwa*; East and South Asia	AFLP	Yashiro et al. (2005)
Molecular	*momordica*; India	RAPD, SSR, and resistance traits	Dhillon et al. (2007)
Molecular and morphological	*cantalupensis*, *inodorus*, *flexousous*, and *reticulatus*	RAPD	Feyzian et al. (2007)
Molecular	*cantalupensis*, *inodorus*, *momordica*, *flexuosus*, *conomon*, *dudaim*, *inodorus x agrestis*; worldwide	SSR	Fukino et al. (2007)
Molecular	*Inodorus*, *conomon*, *dudaim momordica*, *agrestis cantalupensis*, *flexuosus*; Türkiye	RAPD	Sensoy et al. (2007)
Molecular	*makuwa* and *conomon;* South and East Asia	RAPD	Tanaka et al. (2007)
Molecular	*flexuosus*, *momordica*, *chinensis*, *conomon*, *cantalupensis*, and *inodorus*; worldwide	SSR	Fernández-Silva et al. (2008)
Molecular	China, USA, and India	SSR	Kohpayegani and Behbahani (2008)
Molecular	China	RAPD	Luan et al. (2008)
Molecular	*momordica*; worldwide	SSR	Dhillon et al. (2009)
Molecular	Ukraine	AFLP	Nimmakayala et al. (2009)
Molecular	*inodorus*, *cantalupensis*, *flexuosus*, *chinensis*, and *momordica*; worldwide	SSR	Tzitzikas et al. (2009)
Molecular	*cantalupensis*, *conomon*, *momordica*, and *agrestis*; Myanmar	RAPD	Yi et al. (2009)
Molecular		SRAP	Chen et al. (2010)
Molecular and morphological	*inodorus*, *cantalupensis*, *reticulatus*, *ameri*, and *dudaim*; Vietnam	RAPD	Nhi et al. (2010)
Molecular and morphological	*flexuosus* and *dudaim;* Iran	RAPD; fruit traits	Soltani et al. (2010)
Morphological	*reticulatus*, *inodorus*, and *cantalupensis*; Hungary and Türkiye	Flower, plant, fruit, and seed traits	Szamosi et al. (2010)

Molecular	China, Central and South Asia	RAPD and SSR	Aierken et al. (2011)
Molecular	*inodorus*, *ameri*, *makuwa*, *chinensis*, *momordica*, *agrestis*, and *flexuosus*; Southern India	SSR, morphological, and resistance traits	Fergany et al. (2011)
Molecular	*inodorus*; Italy	ISSR	Sestili et al. (2011)
Molecular	*inodorus*, *cantalupensis*, *conomon*, *flexuosus*, *dudaim*, and *momordica*; Türkiye	ISSR, RAPD, and SRAP	Yildiz et al. (2011)
Molecular	Brazil	RAPD	Dantas et al. (2012)
Molecular	*agrestis*, *flexuosus*, *dudaim*, and *conomon*; China	SSR	Gao et al. (2012)
Molecular	*inodorus*, *cantalupensis*, and *flexuosus*; Spain	SSR	Escribano et al. (2012)
Molecular	*inodorus*, *acidulus*, *chandalak*, *chate*, *flexuosus*, *momordica*, *tibish*, and *cantalupensis*; Türkiye	SSR	Kacar et al. (2012)
Molecular		SSR	Liu et al. (2012)
Molecular	Worldwide	SSR	Serres-Giardi and Dogimont (2012)
Morphological	*Cucumis melo* L., *Citrullus lanatus*, *Cucumeropsis mannii*, and *Cucurbita moschata*; Nigeria	Flower and seed traits	Ajuru and Okoli (2013)
Molecular	*inodous*; Türkiye	AFLP	Frary et al. (2013)
Molecular	*inodorus*, *ameri*, *cantalupensis*, *flexuosus*, *dudaim*, *momordica*, *conomon*, *acidulus*, *tibish*, *chito*, and *agrestis*; worldwide	SNP	Esteras et al. (2013)
Morphological	*inodorus*, *dudaim*, and *reticulatus*; Tunisia	Flower, fruit, and seed	Trimech et al. (2013)
Molecular	*agrestis;* Northern and Central China	SSR	Hu et al. (2014)
Molecular	*cantalupensis* and *momordica*; Northern India	SSR and morphological traits	Malik et al. (2014)
Molecular	*inodorus*, *dudaim*, and *cantalupensis*; Iran	SSR	Raghami et al. (2014)
Molecular and Morphological	*flexuosus*, *agrestis*, and *reticulatus*; Türkiye	ISSR; Fruit and seed traits	Yildiz et al. (2014)
Molecular	*flexuosus*; Tunisia	SSR	Henane et al. (2015)
Molecular	*momordica*; India	ISSR	Singh et al. (2015)
Molecular	*flexuosus*; Türkiye	SSR	Solmaz et al. (2016)

(*Continued*)

TABLE 2.1 *(Continued)*
Morphological and Molecular Markers Used to Determine Genetic Diversity in Melons

Diversity	Group/geographic origin	Marker	References
Molecular	India	SSR and resistance loci	Reddy Lakshamana et al. (2016)
Molecular	*inodorus*; Bangladesh	SSR	Molla et al. (2017)
Molecular	*inodorus* and wild melon; Eastern China	SSR	Zhang et al. (2017)
Morphological	Türkiye	Leaf, plant, and fruit traits	Atalmis and Ilbi (2018)
Molecular and morphological	Türkiye	ISSR; plant and fruit traits	Guliyev et al. (2018)
Molecular	*conomon*	SSR	Li et al. (2018)
Molecular and morphological	*flexuosus*; Israel and Palestine	SNP (Seq-DarT array technology)	Omari et al. (2018)
Morphological	Türkiye	68 UPOV traits	Sari and Solmaz (2018)
Molecular	*agrestis*; worldwide		Chomicki et al. (2019)
Molecular	*agrestis*; Asia	SSR	Hu et al. (2019)
Molecular	*agrestis*; worldwide	SNP	Zhao et al. (2019)
Molecular	*flexuosus*; Jordan	AFLP and ISSR	Akash et al. (2020)
Molecular	*flexuosus*; Egypt	ISSR	Ibrahim et al. (2020)
Molecular and morphological	*inodorus*, *flexuosus*, *chate*, and *cantalupensis*; Tunisia	SSR; leaf, stem, flower, fruit, and seed characteristics	Chikh-Rouhou et al. (2021)
Morphological	*kachri*, *momordica*, *conomon*, and *flexuosus*; India	Flower and fruit	Pandey et al. (2021)
Molecular	Northern Cyprus and Türkiye	SRAP	Yilmaz et al. (2021)
Molecular	*agrestis*; Brazil	ISSR and RAPD	Amorim et al. (2022)
Molecular	*flexuosus*, *chate*, and *ibericus*; Spain	SNP	Flores-Leon et al. (2022)
Molecular	China	SSR	Xie et al., 2022

Molecular	*ameri*, *cantalupensis*, *chandalak*, *dudaim*, and *inodorus*; Iraq	URP, SRAP, and CDDP	Aziz and Tahir (2023)
Molecular	Türkiye	ISSR and RAPD	Erdinc et al. (2013)
Molecular	Japan		Shigita et al. (2023)
Molecular	Kazakhstan	RAPD and SSR	Tanaka et al. (2023)
Molecular	*agrestis*; China	SSR and SNP	Zhang et al. (2023)
Molecular	Indonesia	SSR	Adiredjo and Ajidin (2024)
Molecular and morphological	*cantalupensis*; Afghanistan and Iran	ISSR; Fruit characterization	Haqmal et al. (2024)
Morphological traits	Bangladesh	34 agro-morphological traits	Jahan et al. (2024)
Molecular	*conomon*, *makuwa*, *agrestis*, *cantalupensis*, and *inodorus*; Cambodia, Asia	RAPD and SSR	Naznin et al. (2024)
Molecular	India	SSR	Tara et al. (2024)

found in most investigated collections, especially among Indian landraces, indicating that India is a major center of diversification for melon (McCreight, 2004; Fergany et al., 2011; Reddy Lakshamara et al., 2016; Zhao et al., 2019). Efforts are still needed to collect and characterize traditional landraces before the generalization of advanced breeding material and hybrids. Wild melons, traditionally named *agrestis* but corresponding to at least two divergent groups (African/Asian), are still underrepresented in most gene banks and still need to be explored. Oceanian and Australian melons would be of great interest for collecting.

There is an increasing demand for high-yielding, good-quality, and stress-, disease-, and pest-resistant varieties in melons. The available genetic diversity helps plant breeders and researchers overcome the challenges related to these requirements. For this aim, efficient conservation of genetic resources and a better understanding of genetic diversity between and within the *Cucumis* genus are a necessity and crucial for breeders.

2.6 CONCLUSION

Melon is a species within the Cucurbitaceae family with a wide diversity that has been cultivated in almost every geography of the world for many years. Biodiversity and plant genetic resources are the most important treasures in terms of developing new varieties adapted to biotic and abiotic conditions encountered today. Many melon genetic resources are protected and backed up in the genetic resource protection centers/gene banks established in various countries and in the Svalbard Global Seed Center in controlled conditions. Currently, the morphological and genetic characterizations of melon genetic resources in some countries or gene banks are still not complete, and their resistance/tolerance levels against diseases and abiotic stress factors have not been completed. It is of great importance to continue these studies, deposit new seeds in the gene bank, and use new materials for breeding. In addition, it is important to continue taxonomic studies, which are still unclear, to classify new species and better know the domestication processes, which have led to the actual diversity.

REFERENCES

Adiredjo, A. L., & Ajidin, R. B. P. (2024). Genetic diversity and molecular analysis of hybrid melon (*Cucumis melo* L.) genotype based on simple sequence repeats (SSR) markers. In *2023 Brawijaya International Conference (BIC 2023)* (pp. 211–219). Atlantis Press.

Agaoua, A., Rittener, V., Troadec, C., Desbiez, C., Bendahmane, A., Moquet, F., & Dogimont, C. (2022). A single substitution in Vacuolar protein sorting 4 is responsible for resistance to Watermelon mosaic virus in melon. *Journal of Experimental Botany*, 4008–4021. https://doi.org/10.1093/jxb/erac135. (hal-03671982v1)

Aggelis, A., John, I., & Grierson, D. (1997). Analysis of physiological and molecular changes in melon (*Cucumis melo* L.) varieties with different rates of ripening. *Journal of Experimental Botany*, 48, 769–778. https://doi.org/10.1093/jxb/48.3.769

Aierken, Y., Akashi, Y., Nhi, P. T. P., Halidan, Y., Tanaka, K., Long, B.,. . .& Kato, K. (2011). Molecular analysis of the genetic diversity of Chinese Hami melon and its relationship to the melon germplasm from Central and South Asia. *Journal of the Japanese Society for Horticultural Science*, 80(1), 52–65. https://doi.org/10.2503/jjshs1.80.52

Ajuru, M. G., & Okoli, B. E. (2013). The morphological characterization of the melon species in the family *Cucurbitaceae* Juss., and their utilization in Nigeria. *International Journal of Modern Botany*, 3(2), 15–19. https://doi.org/10.5923/j.ijmb.20130302.01

Akash, M., Awad, N., & Kasrawi, M. (2020). Genetic diversity among snake melon landraces (*Cucumis melo* var. *flexuosus*) using molecular descriptors. *Plant Biosystems-An International Journal Dealing with all Aspects of Plant Biology*, 154(2), 206–212. https://doi.org/10.1080/11263504.2019.1587536

Akashi, Y., Fukuda, N., Wako, T., Masuda, M., Kato, K. (2002) Genetic variation and phylogenetic relationships in East and South Asian melons, *Cucumis melo* L., based on the analysis of five isozymes. *Euphytica*, 125, 385–396. https://doi.org/10.1023/A:1016086206423

Amorim, C. C. D., Silva, F. D. D., Bordallo, P. D. N., Aragão, F. A. S. D., & Queiróz, M. A. D. (2022). Genetic divergence between sub-accessions of the melon using molecular markers. *Revista Ciência Agronômica*, 53, e20218007. https://doi.org/10.5935/1806-6690.20220027

Anglin, N. L., Amri, A., Kehel, Z., & Ellis, D (2018). A case of need: Linking traits to genebank accesions. *Biopreservation and Biobanking*, 16(5). https://doi.org/10.1089/bio.2018.0033

Atalmis, F., & Ilbi, H. (2018). Determination of genetic variation among melon genotypes grown in Aegean Region by morphologic and molecular markers. In *XXX International Horticultural Congress IHC2018: V International Symposium on Plant Genetic Resources and International 1297* (pp. 347–358). Acta Horticulturae,

Aziz, R. R., & Tahir, N. A. R. (2023). Genetic diversity and structure analysis of melon (*Cucumis melo* L.) genotypes using URP, SRAP, and CDDP markers. *Genetic Resources and Crop Evolution*, 70(3), 799–813. https://doi.org/10.1007/s10722-022-01462-y

Bagheriyan, S., Karimi, H. R., & Esmaelizadeh, M. (2015). Evaluation of genetic relationships among melon genotypes based on morphological markers. *International Journal of Vegetable Science*, 21(1), 36–52. https://doi.org/10.1080/19315260.2013.818608

Burger, Y., Paris, H. S., Cohen, R., Katzir, N. (2010). Genetic diversity of *Cucumis melo*. *Horticultural Review*, 36, 164–199.

Chen, Y., Li, G., Wang, X. L. (2010) Genetic diversity of a germplasm collection of *Cucumis melo* L. using SRAP markers. *Hereditas*, 32, 744–751. https://doi.org/10.3724/SP.J.1005.2010.00744

Chikh-Rouhou, H., Mezghani, N., Mnasri, S., Mezghani, N., & Garcés-Claver, A. (2021). Assessing the genetic diversity and population structure of a Tunisian melon (*Cucumis melo* L.) collection using phenotypic traits and SSR molecular markers. *Agronomy*, 11(6), 1121. https://doi.org/10.3390/agronomy11061121

Chomicki, G., Schaefer, H., & Renner, S. S. (2019). Origin and domestication of Cucurbitaceae crops: Insights from phylogenies, genomics and archaeology. *New Physiologist*, 226, 1240–1255. https://doi.org/10.1111/nph.16015

Dantas, A. C. D. A., Holanda, I. S. A., Esteras, C., Nunes, G. H. S., & Pico, M. B. (2015). Diversity of melon accessions from northeastern Brazil and their relationships with germplasms of diverse origins. *Journal of the American Society for Horticultural Science*, 140, 504–517. https://doi.org/10.21273/JASHS.140.5.504

Dantas, A. C. D. A., Nunes, G. H. D. S., Araújo, I. S., & Albuquerque, L. B. (2012). Caracterização molecular de acessos de melão coletados no Nordeste brasileiro. *Revista Brasileira de Fruticultura*, 34, 183–189. https://doi.org/10.1590/S0100-29452012000100025

Decker-Walters, D. S., Chung, S. M., Staub, J. E., Quemada, H. D., & López-Sesé, A. I. (2002). The origin and genetic affinities of wild populations of melon (*Cucumis melo*, *Cucurbitaceae*) in North America. *Plant Systematics and Evolution*, 233, 183–197. https://doi.org/10.1007/s00606-002-0191-3

Dhillon, N. P. S., Ranjana, R., Singh, K., Eduardo, I., Monforte, A. J., Pitrat, M., Dhillon, N. K., & Singh, P. P. (2007). Diversity among landraces of Indian snapmelon (*Cucumis melo* var. *momordica). Genetic Resources and Crop Evolution*, 54(6), 1267–1283. https://doi.org/10.1007/s10722-006-9108-2

Dhillon, N. P. S., Singh, J., Fergany, M., Monforte, A. J., & Sureja, A. K. (2009). Phenotypic and molecular diversity among landraces of snapmelon (*Cucumis melo* var. *momordica*) adapted to the hot and humid tropics of eastern India. *Plant Genetic Resources*, 7(3), 291–300. https://doi.org/10.1017/S1479262109990050

Díaz, A., Martín-Hernández, A. M., Dolcet-Sanjuan, R., Garcés-Claver, A., Álvarez, J. M., Garcia-Mas, J., Picó, B., & Monforte, A. J. (2017). Quantitative trait loci analysis of melon (*Cucumis melo* L.) domestication-related traits. *Theoretical and Applied Genetics*, 130, 1837–1856. https://doi.org/10.1007/s00122-017-2928-y

El-Tahir, I. M., &Yousif, M. T. (2004). Indigenous melons (*Cucumis melo* L.) in Sudan: A review of their genetic resources and prospects for use as sources of disease and insect resistance. *Plant Genetic Resources Newsletter*, 138, 36–42.

Endl, J., Achigan-Dako, E. G., Pandey, A. K., Monforte, A. J., Pico, B., & Schaefer, H. (2018). Repeated domestication of melon (*Cucumis melo*) in Africa and Asia and a new close relative from India. *American Journal of Botany*, 105, 1662–1671. https://doi.org/10.1002/ajb2.1172

Erdinc, C., Ekincialp, A., Yildiz, M., Kabay, T., Turkmen, O., & Sensoy, S. (2013). Molecular genetic diversity in Lake Van basin melons (*Cucumis melo* L.) based on RAPD and ISSR markers. *Yuzuncu Yıl University Journal of Agricultural Sciences*, 23(3), 264–270.

Escribano, S., & Lazaro, A. (2009). Agro-morphological diversity of Spanish traditional melons (*Cucumis melo* L.) of the Madrid provenance. *Genetic Resources and Crop Evolution*, 56, 481–497. https://doi.org/10.1007/s10722-008-9380-4

Escribano, S., Lazaro, A., Cuevas, H. E., Lopez-Sesé, A. I., & Staub, J. E. (2012). Spanish melons (*Cucumis melo* L.) of the Madrid provenance: A unique germplasm reservoir. *Genetic Resources and Crop Evolution*, 59, 359–373. https://doi.org/10.1007/s10722-011-9687-4

Esteras, C., Formisano, G., Roig, C., Díaz, A., Blanca, J., Garcia-Mas, J., & Picó, B. (2013). SNP genotyping in melons: Genetic variation, population structure, and linkage disequilibrium. *Theoretical and Applied Genetics*, 126, 1285–1303. https://doi.org/10.1007/s00122-013-2053-5

Fallik, E., Alkali-Tuvia, S., Horev, B., Copel, A., Rodov, V., Aharoni, Y., Ulrich, D., & Schulz, H. (2001). Characterization of 'Galia' melon aroma by GC and mass spectrometric sensor measurements after prolonged storage. *Postharvest Biology and Technology*, 22, 85–91. https://doi.org/10.1016/S0925-5214(00)00185-X

FAOSTAT (2022). Food and Agriculture Organization of the United Nations. http://www.fao.org/faostat/en/#data/QC (accessed on 21 November 2024).

Fergany, M., Kaur, B., Monforte, A., Pitrat, M., Rys, C., Lecoq, H., Dhillon, N., & Dhaliwal, S. (2011) Variation in melon (*Cucumis melo*) landraces adapted to the humid tropics of southern India. *Genetic Resources and Crop Evolution*, 58, 225–243. https://doi.org/10.1007/s10722-010-9564-6

Fernández-Silva, I., Eduardo, I., Blanca, J., Esteras, C., Pico, B., Nuez, F.,. . .& Monforte, A. J. (2008). Bin mapping of genomic and EST-derived SSRs in melon (*Cucumis melo* L.). *Theoretical and Applied Genetics*, 118(1), 139–150. https://doi.org/10.1007/s00122-008-0883-3

Feyzian, E., Javaran, M. J., Dehghani, H., & Zamyad, H. (2007). Analysis of the genetic diversity among some of Iranian melon (*Cucumis melo* L.) landraces using morphological and RAPD molecular markers. *JWSS-Isfahan University of Technology*, 11(41), 151–163.

Flores-Leon, A., Moro, C. P., Marti, R., Beltran, J., Rosello, S., Cebolla-Cornejo, J., & Pico, B. (2022). Spanish melon landraces: Revealing useful diversity by genomis, morphological, and metabolomic analysis. *International Journal of Molecular Sciences*, 23(13), 7162. https://doi.org/10.3390/ijms23137162

Frary, A., Sigva, H. O., Tan, A., Taskin, T., Inal, A., Mutlu, S.,. . .& Doganlar, S. (2013). Molecular genetic diversity in the Turkish national melon collection and selection of a preliminary core set. *Journal of the American Society for Horticultural Science*, 138(1), 50–56. https://doi.org/10.21273/JASHS.138.1.50

Fukino, N., Sakata, Y., Kunihisa, M., & Matsumoto, S. (2007). Characterisation of novel simple sequence repeat (SSR) markers for melon (*Cucumis melo* L.) and their use for genotype identification. *The Journal of Horticultural Science and Biotechnology*, 82(2), 330–334. https://doi.org/10.1080/14620316.2007.11512237

Fuller, D. Q., Denham, T., Arroyo-Kalin, M., Lucas, L., Stevens, C. J., Qin L., Allaby, R. G., & Purugganan, M. D. (2014). Convergent evolution and parallelism in plant domestication revealed by an expanding archaeological record. *Proceedings of the National Academy of Sciences of the United States of America*, 111(17), 6147–6152. https://doi.org/10.1073/pnas.1308937110

Fuller, D. Q., Hosoya, L. A., Zheng, Y., & Qin, L. (2010). A contribution to the prehistory of domesticated bottle gourds in Asia: Rind measurements from Jomon Japan and Neolithic Zhejiang, China. *Economic Botany*, 64(3), 260–265.

Gao, P., Ma, H., Luan, F., & Song, H. (2012). DNA fingerprinting of Chinese melon provides evidentiary support of seed quality appraisal. *PLoS One*, 7, 52431. https://doi.org/10.1371/journal.pone.0052431

Garcia, E., Jamilena, M., Álvarez, J. I., Arnedo, T., Oliver, J. L., & Lozano, R. (1998). Genetic relationships among melon breeding lines revealed by RAPD markers and agronomic traits. *Theoretical and Applied Genetics*, 96, 878–888. https://doi.org/10.1007/s001220050815

Garcia-Mas, J., Benjak, A., Sanseverino, W., Bourgeois, M., Mir, G., Gonźalez, V. M., Heńaff, E., Caṁara, F., Cozzuto, L., Lowy, E., Alioto, T., Capella-Gutieŕrez, S., Blancae, J., Cañizares, J., Ziarsolo, P., Gonzalez-Ibeas, D., Rodriǵuez-Moreno, L., Droege, M., Du, L., & Puigdomeṅech, P. (2012). The genome of melon (*Cucumis melo* L.). *Proceedings of the National Academy of Sciences of the United States of America*, 109(29), 11872–11877.

Garcia-Mas, J., Monforte, A. J., & Arus, P. (2004). Phylogenetic relationships among *Cucumis* species based on the ribosomal internal transcribed spacer sequence and microsatellite markers. *Plant Systematics and Evolution*, 248, 191–203. https://doi.org/10.1007/s00606-004-0170-y

Garcia-Mas, J., Oliver, M., Gomez-Paniagua, H., & de Vicente, M. C. (2000). Comparing AFLP, RAPD and RFLP markers for measuring genetic diversity in melon. *Theoretical and Applied Genetics*, 101, 860–864. https://doi.org/10.1007/s001220051553

Goldman, A. (2002). *Melons, for the Passionate Grower*. Artisan, New York.

Grumet, R., McCreight, J. D., McGregor, C., Weng, Y., Mazourek, M., Reitsma, K.,. . .& Fei, Z. (2021). Genetic resources and vulnerabilities of major cucurbit crops. *Genes*, 12(8), 1222. https://doi.org/10.3390/genes12081222

Guliyev, N., Sharifova, S., Ojaghi, J., Abbasov, M., & Akparov, Z. (2018). Genetic diversity among melon (*Cucumis melo* L.) accessions revealed by morphological traits and ISSR markers. *Turkish Journal of Agriculture and Forestry*, 42(6), 393–401. https://doi.org/10.3906/tar-1707-18

Gull, D. D. (1988). *Harvesting and Handling. Muskmelon Production Guide for Florida, G. J. Hochmuth*. Florida Cooperative Extension Service. University of Florida Cooperative Extension Service Institute of Food and Agricultural Sciences, Gainesville, FL.

Haqmal, M., Haghighi, M., Rahimmalek, M., & Hodaei, M. (2024). Investigate the genetic relationships and morphological diversity of some Afghan and Iranian melon cultivars for breeding purposes. *Plant Molecular Biology Reporter*, 1–13. https://doi.org/10.1007/s11105-024-01433-w

Henane, I., Slimane, R. B., & Jebari, H. (2015). SSR-based genetic diversity analysis of Tunisian varieties of melon (*Cucumis melo* L.) and Fakous (*Cucumis melo* var. *flexuosus*). *International Journal of Advanced Research*, 3(3), 727–734.

Hodges, D. M., & Lester, G. E. (2011). *Cucurbits [Cucumber, Melon, Pumpkin, and Squash]. Health-promoting Properties of Fruit and Vegetables*. CAB İnternational, Wallingford, Oxfordshire, UK, 118–134.

Hu, J. B., Gao, L., Xu, Y., Li, Q., Zhu, H., Yang, L.,. . .& Sun, S. (2019). Microsatellite markers reveal genetic diversity and relationships within a melon collection mainly comprising Asian cultivated and wild germplasms. *BioMed Research International*, 2019(1), 7495609. https://doi.org/10.1155/2019/7495609

Hu, J. B., Wang, P. Q., Li, Q., & Su, Y. (2014). Microsatellite analysis of genetic relationships between wild and cultivated melons in Northwest and Central China. *Molecular Biology Reports*, 41(12), 7723–7728. https://doi.org/10.1007/s11033-014-3668-6

Ibrahim, K., Mohamed, F. H., Abo-Zeid, A. A., Elwan, M. W., & Abdel Salam, M. M. (2020). Genetic diversity in Egyptian snake melon accessions as revealed by Inter Simple Sequence Repeat (ISSR) markers. Catrina: *The International Journal of Environmental Sciences*, 22(1), 71–76.

Jahan, N., Ahmed, Q. M., Khatun, M. F., Rahman, S., Hossain, M. G., Islam, M. R., Jahan, N. (2024). Morphological diversity assessment of local muskmelon (*Cucumis melo* L.) germplasm in Bangladesh. *Journal of Agricultural Sciences*, 19, 439–451.

Janick, J. (2007). Plant exploration: From Queen Hatshepsut to Sir Joseph Banks. Workshop on plant exploration for new fruits – past, present, and future held at the ASHA Annual Conference. *HortScience*, 42, 191–196. https://doi.org/10.21273/HORTSCI.42.2.191

Janick, J., Paris, H. S., & Parrish, D. C. (2007). The Cucurbits of Mediterranean antiquity: Identification of taxa from ancient images and descriptions. *Annals of Botany*, 100, 1441–1457. https://doi.org/10.1093/aob/mcm242

Kacar, Y. A., Simsek, O., Solmaz, I., Sari, N., & Yalcin Mendi, Y. (2012). Genetic diversity among melon accessions (*Cucumis melo*) from Turkey based on SSR markers. *Genetics and Molecular Research*, 11(4), 4622–4631. https://doi.org/10.4238/2012.November.29.2

Katzir, N., Danin-Poleg, Y., Tzuri, G., Karchi, Z., et al. (1996). Length polymorphism and homologies of microsatellites in several Cucurbitaceae species. *Theoretical and Applied Genetics*, 93, 1282–1290. https://doi.org/10.1007/BF00223461

Kaushik, P. (2023). Melon nutraceutomics and breeding. In *Compendium of Crop Genome Designing for Nutraceuticals* (pp. 1053–1067). Singapore: Springer Nature Singapore.

Kirkbride, J. H. (1993). *Biostematic Monograph of the Genus Cucumis* (*Cucurbitaceae*). Parkway, Boone, NC.

Kohpayegani, J. A., & Behbahani, M. (2008). Genetic diversity of some populations of Iranian melon using SSR markers. *Biotechnology*, 7, 19–26. https://doi.org/10.3923/biotech.2008.19.26

Lester, G., & Shellie, K. (2004). *Honey Dew Melon. The Commercial Storage of Fruits, Vegetables, and Florist and Nursery Stocks*. http://www. ba. ars. usda/hb66/contents.html (accessed on 6 July 2008).

Li, L., Xu, X., Wu, P., Zhang, G., & Zhang, X. (2018). Establishment of molecular identity cards for *Cucumis melo* cultivars using SSR markers. *HortScience*, 53(2), 138–143. https://doi.org/10.21273/HORTSCI12537–17

Lija, M., & Beevy, S. S. (2021). A review on the diversity of melon. *Plant Science Today*, 8(4), 995–1003.

Linnaeus, C. (1753). *Species Plantarum*, Vol. 1. Laurentius Salvius, Uppsala, Sweden.

Liu, L., Kakihara, F., & Kato, M. (2004). Characterization of six varieties of *Cucumis melo* L. based on morphological and physiological characters, including shelf-life of fruit. *Euphytica*, 135, 305–313.

Liu, L. Z., Chen, Y. L., Chen, Y. Y., Di, W. J., & Zhu, W. M. (2012). Analysis on genetic dversity of melon varieties using SSR markers. *Journal of Plant Genetic Resources*, 13(3), 381–385.

López-Sesé, A. I., Staub, J. E., & Gómez-Guillamón, M. L. (2003). Genetic analysis of Spanish melon (*Cucumis melo* L.) germplasm using a standardized molecular-marker array and geographically diverse reference accessions. *Theoretical and Applied Genetics*, 108, 41–52. https://doi.org/10.1007/s00122-003-1404-z

Luan, F., Delannay, I., & Staub, J. E. (2008). Chinese melon (*Cucumis melo* L.) diversity analyses provide strategies for germplasm curation, genetic improvement, and evidentiary support of domestication patterns. *Euphytica*, 164, 445–461. https://doi.org/10.1007/s10681-008-9699-0

Malik, A. A., Vashisht, V. K., Singh, K., Sharma, A., Singh, D. K., Singh, H., Monforte, A. J., McCreight, J. D., & Dhillon, N. P. S. (2014). Diversity among melon (*Cucumis melo* L.) landraces from the Indo-Gangetic plains of India and their genetic relationship with USA melon cultivars. *Genetic Resources Crop Evolution*, 61(6), 1189–1208. https://doi.org/10.1007/s10722-014-0101-x

McCollum, T. G., Cantliffe, D. J., & Paris, H. S. (1987). Flowering, fruit-set, and fruit-development in birdsnest-type muskmelons. *Journal of the American Society for Horticultural Science*, 112, 161–164.

McCollum, T. G., Huber, D. J., & Cantliffe D. J. (1988). Soluble sugar accumulation and activity of related enzymes during muskmelon fruit-development. *Journal of the American Society for Horticultural Science*, 113, 399–403.

McCreight, J. D. (2004) Isozyme variation in Indian and Chinese melon (*Cucumis melo* L.) germplasm collections. *Journal of the American Society for Horticultural Science*, 129, 811–818.

Merheb, J., Pawełkowicz, M., Branca, F., Bolibok-Bragoszewska, H., Skarzynska, A., Plader, A., & Chalak, L. (2020). Characterization of Lebanese germplasm of snake melon (*Cucumis melo* subsp. *melo* var. *flexuosus*) using morphological traits and SSR Markers. *Agronomy*, 10, 1293. https://doi.org/10.3390/agronomy10091293

Miccolis, V., & Saltveit, M. E. (1991). Morphological and physiological-changes during fruit-growth and maturation of 7 melon cultivars. *Journal of the American Society for Horticultural Science*, 116, 1025–1029.

Mliki, A., Staub, J. E., Zhangyong, S., & Ghorbel, A. (2001). Genetic diversity in melon (*Cucumis melo* L.): An evaluation of African germplasm. *Genetic Resources Crop Evolution*, 48, 587–597.

Mo, S. Y., Im, S. H., Go, G. D., Ann, C. M., & Kim, D. H. (1998). RAPD analysis for genetic diversity of melon species. *Horticultural Science & Technology*, 16(1), 21–24.

Molla, M. R., Ahmed, I., Rahman, S., Hossain, M. A., Salam, M. A., Chowdhury, M. A. Z., & Rohman, M. M. (2017). Genetic diversity among muskmelon (*Cucumis melo* L.) germplasm in Bangladesh as revealed by microsatellite markers. *African Journal of Agricultural Research*, 12(44), 3203–3213. https://doi.org/10.5897/AJAR2017.12617

Monforte, A. J., Garcia-Mas, J., & Arus, P. (2003). Genetic variability in melon based on microsatellite variation. *Plant Breeding*, 122(2), 153–157. https://doi.org/10.1046/j.1439-0523.2003.00848.x

Munger H. M., & Robinson R. W. (1991). Nomenclature of *Cucumis melo* L. *Cucurbit Genetics Cooperative Reports*, 14, 43–44.

Nakata, E., Staub, J. E., López-Sesé, A. I., & Katzir, N. (2005). Genetic diversity of Japanese melon cultivars (*Cucumis melo* L.) as assessed by random amplified polymorphic DNA and simple sequence repeat markers. *Genetic Resources and Crop Evolution*, 52, 405–419. https://doi.org/10.1007/s10722-005-2258-9

Naudin, C. (1859). Essais d'une monographie des especes et des variet es du genre *Cucumis*. *Annales des sciences naturelles, Botanique, series*. 11, 5.

Naznin, P. M., Imoh, O. N., Tanaka, K., Sreynech, O., Shigita, G., Sophea, Y.,. . .& Kato, K. (2024). Analysis of genetic diversity and population structure in Cambodian melon landraces using molecular markers. *Genetic Resources and Crop Evolution*, 71(3), 1067–1083. https://doi.org/10.1007/s10722-023-01677-7

Nhi, P. T. P., Akashi, Y., Hang, T. T. M., Tanaka, K., Aierken, Y., Yamamoto, T.,. . .& Kato, K. (2010). Genetic diversity in Vietnamese melon landraces revealed by the analyses of morphological traits and nuclear and cytoplasmic molecular markers. *Breeding Science*, 60(3), 255–266. https://doi.org/10.1270/jsbbs.60.255

Nimmakayala, P., Tomason, Y. R., Jeong, J., Vajja, G., Levi, A., Gibson, P., & Reddy, U. K. (2009). Molecular diversity in the Ukrainian melon collection as revealed by AFLPs and microsatellites. *Plant Genetic Resources*, 7(2), 127–134. https://doi.org/10.1017/S1479262108098481

Nunez-Palenius, H. G., Gomez-Lim, M., Ochoa-Alejo, N., Grumet, R., Lester, G., & Cantliffe, D. J. (2008). Melon fruits: Genetic diversity, physiology, and biotechnology features. *Critical Reviews in Biotechnology*, 28, 13–55.

Omari, S., Kamenir, Y., Benichou, J. I. C., Pariente, S., Sela, H., & Perl-Treves, R. (2018). Landraces of snake melon, an ancient Middle Eastern crop, reveal extensive morphological and DNA diversity for potential genetic improvement. *BMC Genetics*, 19, 34. https://doi.org/10.1186/s12863-018-0619-6

Pandey, A., Ranjan, P., Ahlawat, S. P., Bhardwaj, R., Dhariwal, O. P., Singh, P. K., Malav, P. K., Harish, G. D., Prabhu, P., & Agrawal, A. (2021). Studies on fruit morphology, nutritional and floral diversity in less-known melons (*Cucumis melo* L.) of India. *Genetic Resources and Crop Evolution*, 68, 1453–1470. https://doi.org/10.1007/s10722-020-01075-3(0123456789().,-volV)(01234567

Pangalo, K. I. (1951). Anadolu'nun bahçe bostan kültürü. In *Türkiye'nin Zirai Bünyesi (Anadolu)*, Eds. Zhukovski, P., Kıpçak, T. C., Nouruzhan, H., & Türkistanlı, S. Ankara: Türkiye Şeker Fabrikaları A.Ş. Neşriyatı No. 20.

Paris, H. (2016). Overview of the origins and history of the five major cucurbit crops: İssues for ancient DNA analysis of archaeological specimens. *Vegetation History and Archaeobotany*, 25, 405–414. https://doi.org/10.1007/s00334-016-0555-1

Pech, J. C., Bouzayen, M., & Latche, A. (2008). Climacteric fruit ripening: Ethylene-dependent and independent regulation of ripening pathways in melon fruit. *Plant Science*, 175, 114–120. https://doi.org/10.1016/j.plantsci.2008.01.003

Pérez-Jordà, G., Peña-Chocarro, L., Fernández, M. G., & Rodríguez, J. C. V. (2017). The beginnings of fruit tree cultivation in the Iberian Peninsula: Plant remains from the city of Huelva (southern Spain). *Vegetation History and Archaeobotany*, 26, 527–538. https://doi.org/10.1007/s00334-017-0610-6

Périn, C., Gomez-Jimenez, M., Hagen, L., Dogimont, C., Pech, J. C., Latche, A., Pitrat, M., Lelievre, J. M. (2002). Molecular and genetic characterization of a nonclimacteric phenotype in melon reveals two loci conferring altered ethylene response in fruit. *Plant Physiology*, 129, 300–309. https://doi.org/10.1104/pp.010613

Pitrat, M. (2017). Melon genetic resources: Phenotypic diversity and horticultural taxonomy. In *Genetics and Genomics of Cucurbitaceae* (Vol. 3, pp. 25–60), Eds. Grumet, R., Katzir, N., & Garcia-Mas, J, Cham, Switzerland: Springer.

Qi, J., Liu, X., Shen, D., Miao, H., Xie, B., Li, X., Zeng, P., Wang, S., Shang, Y., Gu, X., Du, Y., Li, Y., Lin, T., Yuan, J., Yang, X., Chen, J., Chen, H., Xiong, X., Huang, K., Fei, Z., Mao, L., Tian, L., Städler, T., Renner, S. S. . . .& Huang, S. (2013). A genomic variation map provides insights into the genetic basis of cucumber domestication and diversity. *Nature Genetics*, 5, 1510–1515. https://doi.org/10.1038/ng.2801

Qiu, Z., Shang, X., Ferguson, D. K., & Jinag, H. (2016). Archaeobotanical analysis of diverse plant food resources and palaeovegetation at the Zhumucun site a late Neolithic settlement of the Liangzhu Culture in east China. *Quaternary International*, 426, 75–85. https://doi.org/10.1016/j.quaint.2016.01.010

Raghami, M., López-Sesé, A. I., Hasandokht, M. R., Zamani, Z., Moghadam, M. R. F., & Kashi, A. (2014). Genetic diversity among melon accessions from Iran and their relationships with melon germplasm of diverse origins using microsatellite markers. *Plant Systematics and Evolution*, 300, 139–151. https://doi.org/10.1007/s00606-013-0866-y

Rashid, D., Devani, R. S., Rodriguez-Granados, N. Y., Abou-Choucha, F., Troadec, C., Morin, H., Tan, F. Q., Marcel, F., Huang, H. Y., Hanique, M., Zhang, S., Verdenaud, M., Pichot, C., Rittener, V., Huang, Y., Benhamed, M., Dogimont, C., Boualem, A., & Bendahmane, A. (2023). Ethylene produced in carpel primordia controls *CmHB40* expression to inhibit stamen development. *Nature Plants*, 9, 1675-1687. https://doi.org/10.1038/s41477-023-01511-z

Reddy, Lakshamana D. C., Venkat, S. K., Reddy, A. C., Aswath, C., Avinash, K. N., Nandini, H., & Rao, E. S. (2016). Genetic diversity and population structure of Indian melon (*Cucumis melo* L.) landraces with special reference to disease and insect resistance loci. *Plant Breeding*, 135(3), 384–390. https://doi.org/10.1111/pbr.12356

Renner, S. S., Wu, S., Pérez-Escobar, O. A., Silber, M. V., Fei, Z. J., & Chomicki, G. (2021). A chromosome-level genome of a Kordofan melon illuminates the origin of domesticated watermelons. *PNAS*, 118, 23. https://doi.org/10.1073/pnas.2101486118

Ritschel, P. S., Lins, T. C. D. L., Tristan, R. L., Buso, G. S. C., Buso, J. A., & Ferreira, M. E. (2004). Development of microsatellite markers from an enriched genomic library for genetic analysis of melon (*Cucumis melo* L.). *BMC Plant Biology*, 4, 1–14.

Rolim, P. M., Seabra, L. M. A. J., & de Macedo, G. R. (2020). Melon by-products: Biopotential in human health and food processing. *Food Reviews International*, 36(1), 15–38. https://doi.org/10.1080/87559129.2019.1613662

Sakata, Y., Sugiyama, M. (2002). Characteristics of Japanese cucurbits. In *Second International Symposium on Cucurbits* (pp. 195–203), Eds. Nishimura, S., Ezura, H., Matsuda, T., & Tazuke, A, Tsukuba, Japan: ISHS.

Saputra, H. E., Syukur, M., Suwarno, W. B., & Sobir, S. (2024). Diversity and morphological characteristics of flowers in *reticulatus*, *inodorus*, and *makuwa* group melon (*Cucumis melo*). *Biodiversitas Journal of Biological Diversity*, 25(7). https://doi.org/10.13057/biodiv/d250735

Sari, N., & Solmaz, I. (2018). *Turkey Watermelon and Melon Genetic Resources* (p. 280). 72 Tasarim Ltd. Şti Press, Ankara. ISBN: 978-605-2353-62-2.

Sebastian, P., Schaefer, H., Telford, I. R., & Renner, S. S. (2010). Cucumber (*Cucumis sativus*) and melon (*C. melo*) have numerous wild relatives in Asia and Australia, and the sister species of melon is from Australia. *Proceedings of the National Academy of Sciences*, 107, 14269–14273. https://doi.org/10.1073/pnas.1005338107

Sensoy, S., Buyukalaca, S., & Abak, K. (2007). Evaluation of genetic diversity in Turkish melons (*Cucumis melo* L.) based on phenotypic characters and RAPD markers. *Genetic Resources and Crop Evolution*, 54(6), 1351-1365. https://doi.org/10.1007/s10722-006-9120-6

Serres-Giardi, L., & Dogimont, C. (2012, October 15–18). How microsatellite diversity helps to understand the domestication history of melon. In *Cucurbitaceae 2012, Proceedings of the Xth EUCARPIA Meeting on Genetics and Breeding of Cucurbitaceae* (pp. 254–263), Eds. Sari, Solmaz, & Aras. Antalya, Turkey: Cukurova University Publisher.

Sestili, S., Giardini, A., & Ficcadenti, N. (2011). Genetic diversity among Italian melon *inodorus* (*Cucumis melo* L.) germplasm revealed by ISSR analysis and agronomic traits. *Plant Genetic Resources*, 9(2), 214–217. https://doi.org/10.1017/S1479262111000335

Seungbum, L., Ik, J., Do, Yoon, H., Jung-Ro, L., Seong-Hoon, K., Eunae, Y., Sookyeong, L., Gyu-Taek, C., & Kyung, J. L. (2020). Evaluation of morphological traits and genetic composition in melon germplasm. *Korean Journal of Crop Science*, 65, 485–498. https://doi.org/10.7740/kjcs.2020.65.4.485

Shahwar, D., Khan, Z., & Park, Y. (2023). Molecular marker-assisted mapping, candidate gene identification, and breeding in melon (*Cucumis melo* L.): A review. *International Journal of Molecular Sciences*, 24(20), 15490. https://doi.org/10.3390/ijms242015490

Shellie, K. C., & Lester, G. E. (2002). *Netted Melons. The Comercial Storage of Fruits, Vegetables, and Florist and Nursery Crops* (pp. 1–4), Gross, K. C., Wang, C. Y., & Saltveit, M. USDA Handbook 66, http://www.ba.ars.usda.gov/hb66/(open in a new window)

Shigita, G., Dung, T. P., Pervin, M. N., Duong, T. T., Imoh, O. N., Monden, Y., Nishida, H., Tanaka, K., Sugiyama, M., Kawazu, Y., Tomooka, N., & Kato, K. (2023). Elucidation of genetic variation and population structure of melon genetic resources in the NARO Genebank, and construction of the World Melon Core Collection. *Breeding Science*, 73(3), 269–277. https://doi.org/10.1270/jsbbs.22071

Silberstein, L., Kovalski, I., Huang, R., Anagnostou, K., Jahn, M. M. K., & Perl-Treves, R. (1999). Molecular variation in melon (*Cucumis melo* L.) as revealed by RFLP and RAPD markers. *Scientia Horticulturae*, 79(1–2), 101–111. https://doi.org/10.1016/S0304-4238(98)00199-X

Singh, A. K., Kumar, S., Singh, H., Rai, V. P., Singh, B. D., & Pandey, S. (2015). Genetic diversity in Indian snapmelon (*Cucumis melo* var. *momordica*) accessions revealed by ISSR markers. *Plant Omics*, 8(1), 9–16.

Singh, D., Leskovar, D. I., Sharma, S. P., Sarao, N. K., & Vashisht, V. K. (2020). Genetic diversity and interrelationship among Indian and exotic melons based on fruit morphology, quality components and microsatellite markers. *Physiology and Molecular Biology of Plants*, 26, 985–1002. https://doi.org/10.1007/s12298-020-00814-1

Solmaz, I., Kacar, Y. A., Simsek, O., & Sari, N. (2016). Genetic characterization of Turkish snake melon (*Cucumis melo* L. subsp. *melo flexuosus* Group) accessions revealed by SSR markers. *Biochemical Genetics*, 54, 534–543. https://doi.org/10.1007/s10528-016-9739-8

Solmaz, I., Namli, M., Sari, N., & Abak, K. (2021). *Kavun ıslahı. Sebze Islahı Cilt II (Cucurbits), Türkiye Bitki Islahi Alt Birligi, Gece Kitapligi* (pp. 201–278), Eds. Abak, K., Balkaya, A., Ellialtioglu, S., & Duzyaman, E. Gece Kitapligi,

Soltani, F., Akashi, Y., Kashi, A., Zamani, Z., Mostofi, Y., & Kato, K. (2010). Characterization of Iranian melon landraces of *Cucumis melo* L. Groups Flexuosus and Dudaim by analysis of morphological characters and random amplified polymorphic DNA. *Breeding Science*, 60(1), 34–45. https://doi.org/10.1270/jsbbs.60.34

Staub, J. E., Box, J., Meglic, V., Horejsi, T. F., & McCreight, J. D. (1997). Comparison of isozyme and random amplified polymorphic DNA data for determining intraspecific variation in *Cucumis. Genetic Resources and Crop Evolution*, 44, 257–269.

Staub, J. E., Danin-Poleg, Y., Fazio, G., Horejsi, T., Reis, N., & Katzir, N. (2000). Comparative analysis of cultivated melon groups (*Cucumis melo* L.) using random amplified polymorphic DNA and simple sequence repeat markers. *Euphytica*, 115, 225–241.

Staub, J. E., López-Sesé, A. I., & Fanourakis, N. (2004). Diversity among melon landraces (*Cucumis melo* L.) from Greece and their genetic relationships with other melon germplasm of diverse origins. *Euphytica*, 136, 151–166.

Stepansky, A., Kovalski, I., & Perl-Treves, R. (1999). Intraspecific classification of melons (*Cucumis melo* L.) in view of their phenotypic and molecular variation. *Plant Systematics and Evolution*., 217, 313–332.

Szamosi, C., Solmaz, I., Sari, N., & Barsony, C. (2010). Morphological evaluation and comparison of Hungarian and Turkish melon (*Cucumis melo* L.) germplasm. *Scientia Horticulturae*, 124, 170–182. https://doi.org/10.1016/j.scienta.2009.12.024

Tanaka, K., Nishitani, A., Akashi, Y., Sakata, Y., Nishida, H., Yoshino, H., & Kato, K. (2007). Molecular characterization of South and East Asian melon, *Cucumis melo* L., and the origin of Group *Conomon* var. *makuwa* and var. *conomon* revealed by RAPD analysis. *Euphytica*, 153, 233–247. https://doi.org/10.1007/s10681-006-9259-4

Tanaka, K., Sugiyama, M., Shigita, G., Murakami, R., Duong, T. T., Aierken, Y.,. . .& Kato, K. (2023). Melon diversity on the Silk Road by molecular phylogenetic analysis in Kazakhstan melons. *Breeding Science*, 73(2), 219–229. https://doi.org/10.1270/jsbbs.22030

Tara, K. K., Choudhary, H., Yadav, R. K., Kumari, J., Mishra, D. C., & Kumar, C. (2024). Assessment of genetic diversity and population structure in melon (*Cucumis melo* L.) germplasm using microsatellite markers: Implications towards its varietal improvement. *Indian Journal of Genetics and Plant Breeding*, 84(02), 273–279. https://doi.org/10.31742/ISGPB.84.2.16

Telford, I. R. H., Sebastian, P., Bruhl, J. J., & Renner, S. S. (2011). *Cucumis* (Cucurbitaceae) in Australia and Eastern Malesia, Including Newly Recognized Species and the Sister Species to *C. melo*. *Systematic Botany*, 36, 376–389. https://doi.org/10.1600/036364411X569561

Trimech, R., Zaouali, Y., Boulila, A., Chabchoub, L., Ghezal, I., & Boussaid, M. (2013). Genetic variation in Tunisian melon (*Cucumis melo* L.) germplasm as assessed by morphological traits. *Genetic Resources and Crop Evolution*, 60, 1621–1628. https://doi.org/10.1007/s10722-013-9991-2

Tzitzikas, E. N., Monforte, A. J., Fatihi, A., Kypriotakis, Z., Iacovides, T. A., Ioannides, I. M., & Kalaitzis, P. (2009). Genetic diversity and population structure of traditional Greek and Cypriot melon cultigens (*Cucumis melo* L.) based on simple sequence repeat variability. *HortScience*, 44(7), 1820–1824. https://doi.org/10.21273/HORTSCI.44.7.1820

UPOV (2019). https://www.upov.int/edocs/tgdocs/en/tg104.pdf (accessed on 1 September 2025).

Vishnu-Mittre (1974) The beginnings of agriculture: Palaeobotanical evidence. In *Evolutionary Studies in World Crops* (pp. 3–30), Ed. Hutchinson, J. B, London: Cambridge University Press.

Wojcik, A., Wohlfeiler, J., Huynh, S., Sari, N., Rittener-Ruff, V., Carlos, C., Agaoua, A., Elbelt, S., Lagnel, J., Tamisier, L., Desbiez, C., Boualem, A., Bendahmane, A., Glémin, S., Vigouroux, Y., Tenaillon, M. I., & Dogimont, C. (2024). The selection of new valuable alleles during the melon domestication process in Sudan. In *XIII EUCARPIA Meeting on Cucurbit Genetics and Breeding, 3–6 November 2024*, VicoEquense (NA), Italy,

Xie, H. Y., Gao, C. M., Ye, Y. F., Qin, S. H., Li, G. F., Huang, J. Y., He, Y., Li, T. Y., Liu, T. J., & Hong, R. X. (2022). Genetic diversity analysi of 133 pellice melon germplasm resources based on TP-M13-SSR molecular marker technology. *Journal of Southern Agriculture*, 53(9), 2547–2556. https://doi.org/10.3969/j.issn.2095-1191.2022.09.017

Yashiro, K., Iwata, H., Akashi, Y., Tomita, K., Kuzuya M., Tsumura, Y., Kato, K. (2005). Genetic relationship among East and South Asian melon (*Cucumis melo* L.) revealed by AFLP analysis. *Breeding Science*, 55, 197–206. https://doi.org/10.1270/jsbbs.55.197

Yilmaz, N., Kaya, N., Pinar, H., Hanci, F., & Uzun, A. (2021). Detailed morphological and molecular characterizations of melon (*Cucumis melo* L.) accessions collected from Northern Cyprus and Turkey. *Horticultural Science and Technology*, 39, 471–481.

Yi, S. S., Akashi, Y., Tanaka, K., Cho, T. T., Khaing, M. T., Yoshino, H.,. . .& Kato, K. (2009). Molecular analysis of genetic diversity in melon landraces (*Cucumis melo* L.) from Myanmar and their relationship with melon germplasm from East and South Asia. *Genetic Resources and Crop Evolution*, 56, 1149-1161. https://doi.org/10.1007/s10722-009-9438-y

Yildiz, M., Akgul, N., & Sensoy, S. (2014). Morphological and molecular characterization of Turkish landraces of *Cucumis melo* L. *Notulae Botanicae Horti Agrobotanici Cluj-Napoca*, 42(1), 51–58.

Yildiz, M., Ekbic, E., Keles, D., Sensoy, S., & Abak, K. (2011). Use of ISSR, SRAP, and RAPD markers to assess genetic diversity in Turkish melons. *Scientia Horticulturae*, 130(1), 349–353. https://doi.org/10.1016/j.scienta.2011.06.048

Zhang, J., Yang, J., Lv, Y., Zhang, X., Xia, C., Zhao, H., & Wen, C. (2023). Genetic diversity analysis and variety identification using SSR and SNP markers in melon. *BMC Plant Biology*, 23(1), 39. https://doi.org/10.1186/s12870-023-04056-7

Zhang, Y., Fan, X., Aierken, Y., Ma, X., Yi, H., & Wu, M. (2017). Genetic diversity of melon landraces (*Cucumis melo* L.) in the Xinjiang Uygur autonomous region on the basis of simple sequence repeat markers. *Genetic Resources and Crop Evolution*, 64, 1023-1035. https://doi.org/10.1007/s10722-016-0422-z

Zhao, G., Lian, Q., Zhang, Z. H., Fu, Q. S., He, Y. H., Ma, S., Ruggieri, V., Monforte, A. J., Wang, P. Y., Julca, I., Wang H. S., Liu, J. P., Xu Y., Wang, R. Z., Ji, J. B., Xu, Z. H., Kong, W. H., Zhong, Y., Shang, J. L., Pereira, L., Argyris, J., Zhang, J., Mayobre, C., Pujol, M., Oren, E., Out, D., Wang, J. M., Sun, D. X., Zhao, S. J., Zhu, Y. C., Li, N., Katzir, N., Gur, A., Dogimont, C., Schaefer, H., Fan, W., Bendahmane, A., Fei, Z., Pitrat, M., Gabaldon, T., Lin, T., Garcia-Mas, J., Xu, Y. Y., & Huang, S. W. V. (2019). A comprehensive genome variation map of melon identifies multiple domestication events and loci influencing agronomic traits. *Nature Genetics*, 51, 1607-1615. https://doi.org/10.1038/s41588-019-0522-8

Zheng, X. Y., & Wolff, D. W. (2000). Ethylene production, shelf-life and evidence of RFLP polymorphisms linked to ethylene genes in melon (*Cucumis melo* L.). *Theoretical and Applied Genetics*, 101, 613–624.

Zohary, D., & Hopf, M. (2000) *Domestication of Plants in the Old World*. Clarendon Press, Oxford.

3 Melon Genomics

Lingli Tang and Guangwei Zhao

Melon (*Cucumis melo* L.), an eudicotyledonous diploid species ($2n = 2x = 24$) in the Cucurbitaceae family, holds significant biological and economic importance. As a model organism, it provides unique insights into critical biological processes, such as fruit ripening mechanisms, sex determination pathways, and phloem physiology. Global melon production reached approximately 35 million metric tons in 2023 (FAOSTAT, http://faostat.fao.org), with China, Türkiye, Spain, and the United States emerging as principal producers. This substantial output underscores melons' critical role in agricultural economies worldwide. The species' evolutionary history reveals three independent domestication events – two originating in the Indian subcontinent and one in Africa (Zhao et al., 2019). These distinct domestications resulted in the divergence of two subspecies (*melo* and *agrestis*), each exhibiting unique phenotypic characteristics shaped by selective breeding. Notably, melon's remarkable phenotypic diversity historically informed early genetic theories that laid the foundation for modern plant genetics. Despite serving as a key system for studying diverse biological phenomena and maintaining substantial intraspecific variation, genomic investigations have thus far identified only a limited number of genes associated with crucial agronomic traits. Key characteristics, such as monoecy, flesh pigmentation, and rind coloration, remain partially characterized at the molecular level, highlighting the need for comprehensive functional genomics studies.

3.1 THE GENOMICS AND GENETIC MAPS IN MELON

3.1.1 Genome Assembly Advancements in Melon

The establishment of a reference genome constitutes the cornerstone for genomic research and agricultural utilization of cucurbit species. The first melon reference genome (v3.5) was published by Garcia-Mas et al., 2012) and generated from a doubled-haploid line DHL92 (375 Mb assembly size, 27,427 predicted genes) through interspecific hybridization between divergent *melo* and *agrestis* cultivars. However, its limited scaffold orientation rate (80.8%) hindered practical applications. Subsequent improvements via single nucleotide polymorphism (SNP)–based anchoring (V 3.5.1) significantly enhanced assembly continuity (Argyris et al., 2015). A major upgrade emerged in 2018 with the v3.6 assembly and v4.0 annotation, correcting scaffold orientations and predicting 8,000 novel genes (Ruggieri et al., 2018). This foundation was further strengthened by PacBio SMRT sequencing (v4.0 enhanced), resolving repetitive regions and enabling discrimination between genic and transposable elements (TEs) linked to phenotypic variations (Castanera et al., 2020). Concurrently, de novo assemblies of diverse accessions revealed structural variations (SVs), including

DOI: 10.1201/9781003715757-3

prevalent copy number variations (CNVs) and presence/absence variations (PAVs), with TEs contributing substantially to genomic plasticity (Sanseverino et al., 2015; Saxena et al., 2014; Shin et al., 2019; Yang et al., 2020; Zhang et al., 2019).

Yanno et al. presented a high-quality genome assembly of the semi-climacteric melon cultivar Harukei-3, generated using Oxford Nanopore sequencing technology, which comprised 378 Mb with 33,829 annotated protein-coding genes. This genomic resource was complemented by an updated RNA-seq atlas incorporated into the Melonet-DB database ([Melonet DB] Gene expression atlas), significantly enhancing the platform's utility for melon research (Yano et al., 2020). *Cucumis metuliferus* (African horned cucumber), a wild cucurbit species closely related to cultivated cucumber (*C. sativus*) and melon (*C. melo*), exhibits exceptional resistance to multiple agronomically significant pathogens, including root-knot nematodes and various viral pathogens. A chromosome-level genome was assembled for *C. metuliferus*, revealing a 316-Mb genome that contains 29,039 protein-coding genes with comprehensive annotation. This divergence timeline provides crucial evolutionary context for understanding the development of pathogen resistance mechanisms in this wild species.(Ling et al., 2021). An integrated analysis of multidimensional omics data unveiled extensive genome compartmentalization in *Cucumis melo* of cantaloupe, revealing distinct spatial segregation between constitutive heterochromatin domains and transcriptionally active euchromatin regions. Furthermore, comprehensive cross-varietal genomic comparisons demonstrated progressive evolutionary divergence patterns, with phylogenetic analyses showing the genome has accumulated increasing genetic differentiation from three major melon lineages: Harukei-3 (*C. melo* var. *reticulatus*), Payzawat (*C. melo* var. *inodorus*), and HS (*C. melo* ssp. *agrestis*; Pichot et al., 2022). Otherwise, the wild melon accession P84 genome elucidated the genomic basis of fruit acidity regulation, identifying key SVs affecting *CmPH* gene trajectories during domestication (Chen et al., 2024). Another telomere-resolved assembly (PI511890; 375.0 Mb, contig N50 31.24 Mb) enabled pan-NLRome (leucine-rich repeat receptors, NLRs) analysis across seven accessions, revealing 208 polymorphic and 18 core NLR immune receptors (Mo et al., 2024). Complementary work generated a telomere-to-telomere (T2T) genome (379.2 Mb, N50 31.7 Mb) for landrace Kuizilikjiz, refining 20 meta–quantitative trait loci (QTLs) to facilitate candidate gene discovery (Wei et al., 2023). In *Momordica charantia* var. *abbreviata* (bitter melon), a T2T genome assembly (295.6 Mb, 11 pseudochromosomes) integrated transcriptomic-metabolomic data to unravel cucurbitacin biosynthesis and pigment accumulation mechanisms, enriching Cucurbitaceae phylogenetics (Fu et al., 2023). A chromosome-level T2T genome assembly for 821 (*C. melo ssp. agrestis* var. *acidulus*) was reported, a semi-wild melon with two haplotypes of approximately 373 Mb and approximately 364 Mb, respectively. Comparative genome analysis discovered a significant number of SVs and a copy number variation located in the ToLCNDV-resistance locus on chromosome 11 (Li et al., 2023).

A comparative analysis of chloroplast genomes across nine melon varieties revealed a conserved quadripartite structure, with genome sizes ranging from 155,558 to 156,569 base pairs and GC content between 36.7% and 37%. These organellar genomes contained 127–132 functional genes comprising 85–87 protein-coding sequences, 34–37 tRNA genes, and 6–8 rRNA genes (Hu et al., 2024). Recent pan-genomic investigations have significantly expanded our understanding of melon genetic diversity. A landmark study integrating chromosome-level assembly of wild progenitor *C. melo* ssp. *agrestis* with 23 cultivated accessions identified 3.4 million genetic variations, particularly noting that structural variations in sucrose metabolism pathways underpin key domestication traits (Oren et al., 2022). Molecular dissection further revealed that the *CmPIRL6* locus governs exocarp edibility through PAVs (Lyu et al., 2023). Genome-wide association studies leveraging 297 accessions delineated 709 resistance gene analogs (RGAs), including 603 conserved core genes. Notably, PAV-based analysis identified 13 genomic loci associated with fruit morphology parameters, four residing in pan-genome-specific contigs (Sun et al., 2022). These findings corroborate earlier resequencing efforts that mapped domestication-related selection signatures across melon germplasm (Kishor et al., 2021; Liu et al., 2020; Nimmakayala et al., 2016; Zhao et al., 2019). Collectively establishing a comprehensive framework for decoding agronomic trait in this globally cucurbit is very important. These resources collectively advance comparative genomics and molecular breeding across cucurbit crops.

3.1.2 Genetic Basis of Associated Traits in Melon

Dissecting the genetic basis of agronomic traits constitutes the foundation for germplasm innovation and breeding applications. Over the past decades, substantial progress has been made in melon genetics, with the development of multiple genetic and physical maps alongside comprehensive investigations into key agronomic characteristics, including sex determination, fruit quality, developmental regulation, and disease resistance. The evolution of melon genetic mapping has undergone three distinct phases. Initial mapping efforts employed low-throughput marker systems, including amplified fragment length polymorphisms (AFLPs), random amplified polymorphic DNA (RAPD), and simple sequence repeats (SSRs), resulting in foundational but limited-resolution linkage maps (Baudracco-Arnas and Pitrat, 1996; Gonzalo et al., 2005; Wang et al., 1997). Subsequently, the International Cucurbit Genomics Initiative (ICuGI) established the first consensus linkage map (Diaz et al., 2011), integrating QTL positions for major agricultural traits across studies, although constrained by unmapped ICuGI markers and heterogeneous marker sets across research groups.

The reference melon genome enabled a paradigm shift in mapping strategies (Garcia-Mas et al., 2012). This genomic resource facilitated the precise physical positioning of 1,850 previously ambiguous markers, leading to the development of an integrated genetic-physical framework that has become the current standard reference (Diaz et al., 2015). Contemporary mapping approaches now leverage next-generation sequencing platforms, with multiple studies employing genotyping by sequencing and whole-genome resequencing to construct high-density SNP maps, dramatically improving both resolution and mapping accuracy (Deleu et al., 2009; Lian et al., 2021; Pereira et al., 2018; Wang et al., 2020; Zhang et al., 2020).

3.2 ADVANCES IN THE GENE IDENTIFICATION OF MELON TRAITS

3.2.1 Sex Determination

The molecular regulation of floral sex determination in melon provides critical insights into the developmental control of unisexual flowers. As an established model system for studying sex determination pathways (Latrasse et al., 2017; Nuñez-Palenius et al., 2008), melon exhibits three distinct floral morphotypes: male, female, and bisexual flowers. Plant-level distribution patterns of these floral types define five sexual phenotypes: gynoecy, androecy, hermaphroditism, monoecy, and andromonoecy (Aamir et al., 2021; Martin et al., 2009).

The genetic framework governing this developmental plasticity centers on two key loci: *the andromonoecious* (*a*) gene encoding 1-aminocyclopropane-1-carboxylic acid synthase (*CmACS-7*) and the *gynoecious* (*g*) locus corresponding to *CmWIP1*. The dominant *G* allele exerts carpel suppression through *CmWIP1*-mediated regulation, promoting exclusive male flower formation. Conversely, ethylene biosynthesis mediated by ACC synthase (*CmACS-7*) inhibits stamen development in female flowers. A critical single nucleotide polymorphism in *CmACS-7* governs the monoecy–andromonoecy transition by impairing enzymatic activity, with monoecious (*A-G-*) plants producing male flowers on primary axes versus andromonoecious (*aaG-*) individuals bearing hermaphroditic flowers on lateral branches. Homozygous recessive gynoecious (*AAgg*) and hermaphroditic (*aagg*) genotypes develop exclusively female or bisexual flowers, respectively (Boualem et al., 2008; Martin et al., 2009). The establishment of stable gynoecy in melon requires three critical genetic components: two previously characterized loci and a third recessive m allele in the *CmACS11* gene (Aamir et al., 2021; Boualem et al., 2015). However, melon sex expression demonstrates notable phenotypic plasticity, emerging from complex interplay between hormonal regulation, environmental stimuli, and developmental programming. This dynamic system necessitates three sequential processes: signal integration through molecular sensors, transcriptional reprogramming in response to these signals, and sustained maintenance of acquired states until subsequent environmental shifts (Baulcombe and Dean, 2014).

Epigenetic mechanisms, particularly those involving chromatin modifications, provide crucial molecular memory systems that stabilize these phenotypic commitments. The female-promoting *CmACS11* executes its sexual determination function through H3K27me3-mediated transcriptional silencing of the male-promoting *CmWIP1*. A tripartite regulatory network emerges where *ACS7*, *ACS11*, and *WIP1* interact through epistatic/hypostatic relationships, with their functional states being epigenetically modulated by histone modification signatures (H3K9ac and H3K27me3; Aamir et al., 2021; Latrasse et al., 2017). Recent genetic dissection of floral mutants identified a Harbinger transposon insertion disrupting *CmEIN2* (ethylene insensitive 2), causing a bisexual-to-male transition. Transcriptomic analyses position *CmEIN2* as a dual-function regulator: upon *CmACS11* activation, it represses the carpel inhibitor CmWIP1 while coordinating ethylene-mediated sex determination pathways (Huang et al., 2024). QTL mapping localized *Opbf3.1* (containing *CmACS11*) and *tpbf8.1* as critical determinants of hermaphroditism in UT1 plants, potentially through inhibitory effects on *CmWIP1* and *CmACS-7* functions, respectively (Nashiki et al., 2023). Three non-synonymous SNPs in *CmACS11* differentiate parental lines drive phenotypic differences. Evolutionary analyses of melon germplasm identified an African-origin haplotype influencing EIN3/EIL1-*CmHB40* promoter interactions associated with bisexual flower development (Rashid et al., 2023). These findings collectively illustrate how epigenetic memory, hormonal signaling, and genetic architecture integrate with environmental cues to modulate sex determination plasticity in melon.

3.2.2 Soluble Solid Content

Fruit sensory quality is primarily governed by the interplay of sugar composition, organic acid levels, and volatile aromatic compounds. In melon, sugar content serves as not only a critical determinant of fruit quality and consumer preference but also a principal target for breeding improvement programs (Argyris et al., 2017). This preferential sucrose accumulation highlights the importance of understanding its molecular regulatory mechanisms to facilitate genetic enhancement strategies. Current research reveals that sucrose metabolism involves a complex enzymatic network comprising nearly 20 enzymes, particularly invertases (INVs), sucrose synthase (SS), and sucrose phosphate synthase (SPS). The equilibrium between sucrose biosynthesis (mediated by SPS) and catabolism (controlled by INVs and SS) ultimately dictates sucrose accumulation levels. Genomic analyses have identified 42 melon genes associated with sugar metabolic pathways, though functional validation remains limited (Dai et al., 2011). There has been some progress in studying the genetic control of sugar accumulation. Burger et al. considered that a single major gene, *suc*, controls sucrose accumulation (Burger et al., 2002). Initial reports proposed monogenic control through the *suc* locus, whereas subsequent QTL mapping identified six major genomic regions influencing sugar content across chromosomes 2, 3, 5, and 8 (Diaz et al., 2011; Harel-Beja et al., 2010). Recent meta-analyses integrating 18 mapping studies established consensus QTLs for Brix value and sugar components, with particular interest in *SUCQSC5.1* – a major QTL on chromosome 5 affecting soluble solids and sucrose content, where *MELO3C014519* emerges as a promising candidate gene (Argyris et al., 2017).

Emerging evidence underscores the dual importance of sugar synthesis and translocation mechanisms. While mesophyll cells initiate sugar production, efficient transport to sink organs (particularly fruits) proves equally critical. Among three identified tonoplast sugar transporters (*CmTST1–3*), *CmTST2* demonstrates predominant involvement in fruit sugar accumulation (Cheng et al., 2018; Schemberger et al., 2020). Transcriptomic analyses further implicate specific isoforms – notably *CmACLb1* showing strong positive correlation with sucrose levels – as key metabolic regulators (Ren et al., 2023). Recent advancements in multi-environment QTL mapping have identified stable genomic regions governing soluble solid content (SSC), including *SSC6.1* and *SSC11.1* loci (Zhao et al., 2023). Notably, chromosome 3 harbors two high-impact QTL clusters (*qSSC3–1* and *qSSC3–2*) explaining >10% phenotypic variance each, co-localized with multiple trait-associated SNPs. Similar significant QTL–SNP associations were detected on chromosomes 4 (*qSSC4–1/4-2*) and 6 (*qSSC6–2*; Yan et al., 2023). This observed QTL's heterogeneity likely stems from both genetic background diversity across studies and the inherent complexity of sugar metabolic pathways, presenting challenges for functional validation of candidate genes. Future research should prioritize integrating multi-omics approaches to elucidate the spatiotemporal regulation of sugar metabolism and validate transport mechanisms in melon fruit development.

3.2.3 Fruit Acidity

Fruit acidity, along with sugar content and volatile compounds, serves as a crucial determinant of organoleptic quality in most fleshy fruits. Notably, melon represents an exceptional case among fleshy fruits, exhibiting remarkably low acidity across most cultivars except for certain acidic varieties (Burger et al., 2003). In plant systems, titratable acidity and pH primarily reflect the accumulation of organic acids, with citric acid constituting the predominant form throughout melon fruit development and demonstrating a strong positive correlation with titratable acidity (Tang et al., 2010). Genetic studies have revealed that melon flesh acidity is regulated by a single dominant locus, such as *PH*, which encodes a transmembrane transporter protein and has been validated through multiple population studies (Burger et al., 2003; Cohen et al., 2012). Interestingly, biochemical analyses have unexpectedly identified substantial concentrations of nucleosides (particularly adenosine) and organic acids (citrate and malate) in melon fruits. Subsequent research indicates that the PH protein, localized to the endoplasmic reticulum, may indirectly modulate acidity through two potential mechanisms: proton gradient regulation or nucleoside/acid metabolism (Cohen et al., 2014). Genomic investigations uncovered a characteristic 12-bp insertion in non-acidic cultivars of the melo group, although this mutation shows incomplete penetrance in the *agrestis* group. Intriguingly, a candidate gene encoding putative citrate synthase (*MELO3C011482*) on chromosome 3 was specifically identified in *agrestis* accessions (Zhao et al., 2019). These findings not only elucidate the genetic basis of melon acidity but also provide compounding evidence supporting the hypothesis of independent domestication pathways for the *melo* and *agrestis* groups.

3.2.4 Peel Color

Rind coloration constitutes a crucial quality attribute in melons that significantly influences consumer preference and marketability. Commercial cultivars primarily exhibit three dominant peel phenotypes, including chlorophyll-mediated green, carotenoid/flavonoid-dependent white, and yellow rinds. The transition from green to alternative coloration patterns arises from differential accumulation of photosynthetic pigments (chlorophylls) and photoprotective pigments (carotenoids and flavonoids) during fruit development (Burger et al., 2006; Feder et al., 2015).

Genetic analyses of F_2 populations from green × yellow crosses reveal that green pigmentation demonstrates dominant epistasis over non-green phenotypes. The *MELO3C003375* locus on chromosome 4 has been identified as a key regulator of chlorophyll accumulation, with functional conservation observed between melon and watermelon pigment metabolism (Oren et al., 2019; Zhao et al., 2019). Beyond basic coloration, certain accessions exhibit secondary patterning features including speckles, stripes, and mottling. Recent studies characterize two epistatic genes (*CmMt1 and CmMt2*) governing chlorophyll distribution in mottled phenotypes and a striped peel determinant (*st3*). Notably, *CmMt1* corresponds to *MELO3C003375*, while *st3* co-localizes to adjacent chromosomal regions, suggesting this locus functions as a pigment accumulation hub (Liu et al., 2019; Shen et al., 2021). Supplementary genetic components include *MELO3C003097* on chromosome 8, encoding the chloroplast development factor SLOW GREEN 1 that potentially modifies rind coloration as a minor QTL (Zhao et al., 2019). Distinct from chlorophyll/carotenoid pathways, 'Canary Yellow' cultivars employ naringenin chalcone as their principal yellow flavonoid pigment. This metabolic route is negatively regulated by *CmKFB* (chromosome 10), encoding a kelch-domain F-box protein that suppresses chalcone accumulation (Feder et al., 2015). Although numerous genes governing rind color have been extensively studied, the underlying genetic mechanisms remain complex and incompletely characterized. Moving from gene discovery to functional validation and ultimately breeding applications represents the next critical frontier in this field.

3.2.5 Flesh Color

Flesh color represents a critical quality trait of melon that significantly influences both consumer preference and nutritional composition. Commercial varieties primarily exhibit three distinct flesh colors – orange, white, and green – determined by two key genetic loci: white flesh (*wf*) and green flesh (*gf*). The orange phenotype is governed by the dominant *Gf* allele, while its recessive counterpart *gf* produces green flesh (Clayberg, 1992). Notably, genotypes with *gfgf* configuration display either green (*Wf-*) or white flesh (*wfwf*), establishing an interesting epistatic relationship between these loci. Recent molecular breakthroughs have clarified the genetic basis of these traits. The carotenoid accumulation gene *CmOr*, identified as identical to the *gf* locus (Tzuri et al., 2015), explains the biochemical mechanism behind orange/green color differentiation. Complementary investigations combining QTL analysis and GWAS revealed a critical 96-kb genomic interval containing 11 protein-coding genes. Intriguingly, while the previously proposed *Wf* candidate gene *MELO3C003069* resides 202 kb outside this interval, *MELO3C003097* has emerged as the prime candidate for the *Wf* locus, supported by multiple studies (Cohen et al., 2014; Galpaz et al., 2018; Zhao et al., 2019). A pivotal discovery involves *Cme13C08g017690*, a gene homologous to *Arabidopsis ETHYLENE-DEPENDENT GRAVITROPISM-DEFICIENT AND YELLOW-GREEN* 1 (EGY1). Functional characterization revealed that a single SNP in its coding region creates a premature stop codon, potentially disrupting normal gene function. This finding was corroborated through genome editing of *CsEGY1*, the cucumber ortholog, which demonstrated its essential role in golden-fruit pigmentation (Li et al., 2025).

3.2.6 Molecular Regulation of Plant Pathogens in Melon

Powdery mildew (PM), a major fungal disease affecting melon cultivation, is predominantly caused by two pathogenic species: *Podosphaera xanthii* (comprising seven physiological races) and *Golovinomyces cichoracearum* (with two races; Kenigsbuch and Cohen, 1989; McCreight et al., 2005; Perchepied et al., 2005). Notably, the Mildew Resistance Locus O (MLO) gene family, characterized by seven conserved transmembrane domains, plays critical roles in plant-pathogen interactions. In melon, 14 MLO family members have been identified, among which *CmMLO5* (*MELO3C012438*) has been demonstrated to negatively regulate PM resistance in the susceptible cultivar 'Top Mark', with functional activity peaking at 24 and 72 hours post-inoculation (Zhang et al., 2023a). Geographic variations in climatic conditions have led to distinct physiological races of PM globally, complicating resistance breeding. To date, multiple QTLs conferring race-specific resistance have been mapped, including *BPm12.1* from MR-1, *PmEdisto47–1* from Edisto47, and *Pm-2F* from K7–1 (Li et al., 2017; Ning et al., 2013; Zhang et al., 2012). However, discrepancies in inheritance patterns across studies likely arise from differential utilization of pathogen races and resistant germplasms.

Recent advancements in melon disease resistance research have yielded significant genetic insights. The DM9.1 marker system has been validated for cloning downy mildew (DM) resistance genes and shows promise as a molecular tool for DM-resistant melon breeding programs (Zhang et al., 2023b). Studies on *Pseudoperonospora cubensis*, the causal agent of cucurbit downy mildew (CDM), reveal its destructive potential through severe necrosis and defoliation in melon. Genetic mapping identified two major QTLs: *qPcub−10.3–10.4* on chromosome 10 demonstrated consistent resistance across experimental conditions, while *qPcub−8.3* on chromosome 8 showed environment-specific expression limited to greenhouse trials (Toporek et al., 2023). Notably, these QTLs occupy different chromosomal positions than previously reported resistance loci (*qPcub−8.2* and *qPcub−10.1*) effective against Clade 1/mating type A2 strains of *P. cubensis*. Genetic resistance to gummy stem blight (GSB) appears to follow a monogenic inheritance pattern across diverse melon cultigens. Five dominant loci (Gsb-1 through Gsb-4 and Gsb-6) and one recessive locus (Gsb-5) have been characterized from various plant introduction (PI) lines: Gsb-1 (PI 140471), Gsb-2 (PI 157082), Gsb-3 (PI 511890), Gsb-4 (PI 482398), and Gsb-5 (PI 482399; Frantz and Jahn, 2003; Hassan et al., 2018; Prasad and Norton, 1967). While *MELO03C012987* emerges as a potential candidate gene, most GSB-resistance mechanisms remain genetically unresolved (Hassan et al., 2018; Hu et al., 2018; Virtuoso et al., 2022). Fusarium wilt (FW)–resistance research has identified five major loci combating different *Fusarium oxysporum* (Fom) races: (1) Fom-2: resistance to races 0 and 1 (Joobeur et al., 2004); (2) Fom-1: race 2 resistance (Tezuka et al., 2009); (3) Fom-4: recessive resistance to races 0 and 2, linked to Fom-1 (Oumouloud et al., 2010); (4) Fom-3: putative allelic variant of Fom-1 with race 0 and 2 resistance (Zink and Gubler, 1985); and (5) Fom-5: novel dominant resistance locus (Deol et al., 2022). Emerging evidence implicates zinc finger proteins (ZFPs) in stress responses, with *CmC3H24* and *CmC3H33* potentially regulating FW resistance through heavy metal stress pathways (Oumouloud and Álvarez, 2016; Zheng et al., 2025). The SUPERMAN (SUP) proteins, which belong to the single C2H2 ZFP subclass, participate in plant morphogenesis. *CmSUP7* was identified in regulating fruit development (Dai et al., 2025). This comprehensive genetic framework provides valuable targets for molecular breeding strategies against major melon pathogens.

Melon also faces significant challenges from several viral pathogens, with cucumber mosaic virus (CMV), zucchini yellow mosaic virus (ZYMV), cucumber green mottle virus (CGMMV), and melon necrotic spot virus (MNSV) being the most prevalent viral diseases (Argyris et al., 2015; Rodriguez-Hernandez et al., 2012; Thakur et al., 2019). To date, approximately 200 viral resistance genes have been identified in melon, with more than 50% of these being recessive genes encoding eukaryotic translation initiation factors (eIFs) that mediate viral resistance (Diaz-Pendon et al.,

2004; Truniger and Aranda, 2009). Notably, melon plants with *Cm-eIF4E* knockdown exhibit broad-spectrum resistance to multiple viruses, including MNSV, ZYMV, cucumber vein yellowing virus (CVYV), and Moroccan watermelon mosaic virus (MWMV; Rodriguez-Hernandez et al., 2012). Research on specific virus resistance mechanisms reveals distinct genetic architectures. The CMV resistance in melon accession PI 161375 involves a single major gene coupled with at least two QTLs (Guiu-Aragonés et al., 2014), while the *CmVPS41* gene has been identified as a key regulator preventing CMV phloem entry (Pascual et al., 2019). For ZYMV resistance, the dominant Zym locus in PI 414723 confers protection through a single allele (Adler-Berke et al., 2021). Among resistance gene classes, nucleotide-binding leucine-rich-repeat (NBS-LRR) genes play a crucial role, with the *Vat* gene demonstrating effectiveness against aphid vectors (Marone et al., 2013). However, the evolutionary arms race between pathogens and host plants presents challenges, as pathogen mutations frequently overcome race-specific resistance conferred by most NB-LRR genes (Tian et al., 2003). These findings highlight the critical need for identifying and cloning broad-spectrum resistance genes, particularly through continued exploration of the Nucleotide-Binding Site (NBS) disease resistance gene family, to develop durable viral resistance in melon cultivars.

3.2.7 Abiotic Stress Responses in Melon

Melon faces significant challenges from abiotic stresses, including salinity, drought, and temperature extremes, which critically impair growth and productivity. Unlike biotic stresses, the genetic basis of abiotic stress tolerance in melon remains underexplored, partly due to the developmental stage-specific sensitivity of crops to environmental stressors. Recent advances, however, have uncovered molecular players and pathways essential for stress adaptation.

Potassium (K^+) homeostasis plays a pivotal role in salinity tolerance. The Shaker-like K^+ outward rectifying (SKOR) channel facilitates long-distance K^+ transport from roots to shoots, with *CmSKOR* overexpression shown to enhance salinity tolerance in *Arabidopsis* by optimizing shoot K^+ allocation (Huang et al., 2018). Additionally, *CmLOX08* and *CmCADs* have been implicated in salt stress responses through pathways involving lipid metabolism and lignin biosynthesis (Liu et al., 2018; Wang et al., 2019). Drought tolerance involves jasmonic acid (JA)–mediated signaling and stomatal regulation. *CmLOX10* promotes JA accumulation and stomatal closure under drought stress (Xing et al., 2020). While *CmPPR4*, a pentatricopeptide repeat protein, and six zinc-finger transcription factors (*CmC3H4*, *CmC3H7*, *CmC3H13*, *CmC3H24*, *CmC3H33*, *CmC3H38*) are strongly associated with drought resistance (Xia et al., 2025; Zheng et al., 2025). Low-temperature tolerance involves abscisic acid (ABA) signaling and redox homeostasis. *CmNCED3*, a key enzyme in ABA biosynthesis, contributes to cold adaptation alongside drought and salt responses (Chevilly et al., 2021). The cytokinin response regulator *CmRR6* and pseudoresponse regulator *CmPRR3* regulate cold stress signaling (Li et al., 2024), while *CmPYL6* functions as a critical ABA receptor in cold tolerance (Liu et al., 2024). Furthermore, *CmTGA8* interacts with antioxidant enzymes (*CmAPX1* and *CmGSTU25*) to enhance cold resilience by boosting glutathione-S-transferase (*GST*) activity and reducing

TABLE 3.1
Genes Underlying Agronomically Important Traits in Melon

Traits	Gene	Reference
Sex determination	*CmACS7*	Rashid et al. (2023)
	CmACS11	Cheng et al. (2018)
	CmWIP1	Ren et al. (2023)
	CmEIN2	Huang et al. (2024)
	EIN3/EIL1-*CmHB40*	Rashid et al. (2023)
Fruit soluble solid content	*CmTST2*	Cheng et al. (2018)
	CmACLb1	Ren et al. (2023)
Fruit acidity	*CmPH*	Cohen et al. (2012)
Fruit peel color	*CmMt1 and CmMt2*	Liu et al. (2019); Shen et al. (2021)
	MELO3C003097	Zhao et al. (2019)
Fruit flesh color	*CmOr*	Tzuri et al. (2015)
	Cme13C08g017690 (*EGY1*)	Li et al. (2025)
Plant pathogen	*CmMLO5*	Zhang et al. (2023a)
	CmSUP7	Dai et al. (2025)
	CmC3H24 and *CmC3H33*	Zheng et al. (2025)
	CmeIF4E	Rodriguez-Hernandez et al. (2012)
	CmVPS41	Pascual et al. (2019)
	CmPIF8-CmERF27-*CmACS10*	Wu et al. (2024)
Abiotic stress	*CmSKOR*	Huang et al. (2018)
	CmLOX08 and *CmCADs*	Liu et al. (2018); Wang et al. (2019)
	CmLOX10	Xing et al. (2020)
	CmNCED3	Chevilly et al. (2021)
	CmWOX13a and *CmWOX13b*	Tang et al. (2023)
	CmPRR3	Liu et al. (2024)
	CmPPR4	Xia et al. (2025)
	CmTGA8	Xu et al. (2025)

oxidative damage (Xu et al., 2025). WUSCHEL-related homeobox (WOX) transcription factors are developmental regulators in plants, particularly *CmWOX13a* and *CmWOX13b* in melon, and are constitutively expressed under drought and cold stresses, linking developmental plasticity to environmental adaptation (Tang et al., 2023).

3.3 BREEDING OF MELON

The field of melon breeding has evolved significantly, driven by shifting priorities and technological advancements. The following is a structured synthesis of key trends, challenges, and future directions in melon breeding.

3.3.1 Development of Breeding Objectives

Historically, the focus has centered on yield, shelf life, and resistance to biotic (e.g., diseases) and abiotic (e.g., drought) stresses. Successful examples include PM-resistant varieties developed using Indian accessions and Turkish accessions resistant to viruses such as ZYMV and WMV (Ekbic et al., 2010). Modern priorities place an increasing emphasis on fruit quality, including internal traits, such as sugar/acid content, flavor, and flesh texture, and external traits, such as size, rind color, and shape, such as the *CR-Cmfull* plant created by CRISPR-Cas9 that shows enhanced longitudinal fruits (Tang et al., 2025). Molecular markers for traits like rind color and acid content enable early selection, but flavor complexity (governed by multiple genes and pathways) remains a major challenge (Moing et al., 2020; Tieman et al., 2017).

3.3.2 Disease Resistance and Genetic Diversity

Diverse germplasm sources reveal that resistant accessions often originate from primary/secondary diversity centers (e.g., *Momordica* and acidulous groups for mildew resistance). Exotic germplasm (PI 161375 for CMV/aphid resistance) serves as a critical resource (Guiu-Aragonés et al., 2014). Subspecies differentiation (*melo* vs. *agrestis*) offers opportunities for heterosis through inter-subspecies crosses, despite low nucleotide diversity within *agrestis* (Zhao et al., 2019).

3.3.3 Fruit Quality and Multi-Omics Approaches

Flavor complexity requires integration of genetics, transcriptomics, and metabolomics. Volatile organic compounds (VOCs) study shows *ETHQB3.5* regulates climacteric ripening-associated VOCs (Dos-Santos et al., 2023), while non-climacteric volatiles are governed by distinct networks. *CmTCP1* and *CmYAB1* synergistically regulate tendril development (Mizuno et al., 2024), and terpene cyclase genes drive sesquiterpenoid/triterpenoid biosynthesis (Ren et al., 2025).

3.3.4 Hybrid Vigor (Heterosis) Strategies

Inter-subspecies crossing *melo* (high diversity) with *agrestis* (low diversity but ecological resilience) can enhance hybrid vigor, addressing traits like stress tolerance and yield (Lippman and Zamir, 2007). The limitation is that minimal heterosis has been observed in intra-*agrestis* crosses due to genetic uniformity.

3.3.5 Integration of Advanced Technologies

Marker-assisted selection accelerates trait introgression but faces limitations in polygenic traits (e.g., flavor). Whole-genome selection leverages resequencing data and SNPs for predicting complex traits. CRISPR/Cas9 technology enables precise editing for *CR-Cmnac-nor* and *CR-Cmaco1* delayed melon fruit ripening and extended the shelf life of melon fruit (Liu et al., 2022; Nonaka et al., 2023).

3.4 CONCLUDING REMARKS

Melon stands as a globally significant horticultural crop characterized by remarkable genetic diversity. Future breeding efforts are expected to prioritize two crucial traits: enhanced resistance to biotic/abiotic stresses and improved fruit quality parameters. This necessitates the identification of novel genetic resources possessing horizontal resistance mechanisms or unique fruit quality attributes. However, these desirable traits are governed by complex genetic architectures involving polygenic control and multi-locus interactions. While QTLs associated with these characteristics have been documented, the molecular basis underlying these associations – particularly the identification of causal genes and key genetic variations – remains poorly elucidated. This persisting knowledge gap could be addressed through comprehensive multi-omics approaches integrating genomics, transcriptomics, metabolomics, and advanced bioinformatics. Current breeding programs predominantly rely on conventional methodologies, which may prove inadequate for addressing modern agricultural challenges. Conversely, emerging biotechnological strategies – including genome-wide selection, transgenic approaches, and CRISPR-based genome editing – present transformative opportunities. Strategic implementation of these tools could revolutionize melon improvement by enabling precise manipulation of fruit quality traits while accelerating breeding cycles.

REFERENCES

Aamir, M., Karmakar, P., Singh, V.K., Kashyap, S.P., Pandey, S., Singh, B.K., Singh, P.M., and Singh, J. (2021). A novel insight into transcriptional and epigenetic regulation underlying sex expression and flower development in melon (*Cucumis melo* L.). *Physiologia Plantarum 173*, 1729–1764.

Adler-Berke, N., Goldenberg, Y., Brotman, Y., Kovalski, I., Gal-On, A., Doniger, T., Harel-Beja, R., Troadec, C., Bendahmane, A., Pitrat, M., *et al.* (2021). The melon zym locus conferring resistance to ZYMV: High resolution mapping and candidate gene identification. *Agronomy 11* (12), 2427.

Argyris, J.M., Díaz, A., Ruggieri, V., Fernández, M., Jahrmann, T., Gibon, Y., Picó, B., Martín-Hernández, A.M., Monforte, A.J., and Garcia-Mas, J. (2017). QTL analyses in multiple populations employed for the fine mapping and identification of candidate genes at a locus affecting sugar accumulation in melon (*Cucumis melo* L.). *Frontiers in Plant Science 8*.

Argyris, J.M., Pujol, M., Martín-Hernández, A.M., and Garcia-Mas, J. (2015). Combined use of genetic and genomics resources to understand virus resistance and fruit quality traits in melon. *Physiologia Plantarum 155*, 4–11.

Baudracco-Arnas, S., and Pitrat, M. (1996). A genetic map of melon (*Cucumis melo* L.) with RFLP, RAPD, isozyme, disease resistance and morphological markers. *Theoretical and Applied Genetics 93*, 57–64.

Baulcombe, D.C., and Dean, C. (2014). Epigenetic regulation in plant responses to the environment. *Cold Spring Harbor Perspectives in Biology 6*, a019471.

Boualem, A., Troadec, C., Camps, C., Lemhemdi, A., Morin, H., Sari, M.A., Fraenkel-Zagouri, R., Kovalski, I., Dogimont, C., and Perl-Treves, R. (2015). A cucurbit androecy gene reveals how unisexual flowers develop and dioecy emerges. *Science 350*, 688–691.

Boualem, F.M., Fernandez, R., Troadec, C., Martin, A., Morin, H., Sari, M.A., Collin, F., Flowers, J.M., Pitrat, M., Purugganan, M.D., *et al.* (2008). A conserved mutation in an ethylene biosynthesis enzyme leads to andromonoecy in melons. *Science 321*(5890), 836–838.

Burger, Y., Sa'ar, U., Distelfeld, A., Katzir, N., Yeselson, Y., Shen, S., and Schaffer, A.A. (2003). Development of sweet melon (*Cucumis melo*) genotypes combining high sucrose and organic acid content. *Journal of the American Society for Horticultural Science 128*, 537–540.

Burger, Y., Saar, U., Katzir, N., and Paris, H.S. (2002). A single recessive gene for sucrose accumulation in *Cucumis melo* fruit. *Journal of the American Society for Horticultural Science 127*, 938–943.

Burger, Y., Sa'ar, U., Paris, H., Lewinsohn, E., Katzir, N., Tadmor, Y., and Schaffer, A. (2006). Genetic variability for valuable fruit quality traits in *Cucumis melo*. *Israel Journal of Plant Sciences 54*, 233–242.

Castanera, R., Ruggieri, V., Pujol, M., Garcia-Mas, J., and Casacuberta, J.M. (2020). An improved melon reference genome with single-molecule sequencing uncovers a recent burst of transposable elements with potential impact on genes. *Frontiers in Plant Science 10*.

Chen, X., Li, H., Dong, Y., Xu, Y., Xu, K., Zhang, Q., Yao, Z., Yu, Q., Zhang, H., and Zhang, Z. (2024). A wild melon reference genome provides novel insights into the domestication of a key gene responsible for melon fruit acidity. *Theoretical and Applied Genetics 137*.

Cheng, J., Wen, S., Xiao, S., Lu, B., Ma, M., and Bie, Z. (2018). Overexpression of the tonoplast sugar transporter *CmTST2* in melon fruit increases sugar accumulation. *Journal of Experimental Botany 69*, 511–523.

Chevilly, S., Dolz-Edo, L., Martínez-Sánchez, G., Morcillo, L., Vilagrosa, A., López-Nicolás, J.M., Blanca, J., Yenush, L., and Mulet, J.M. (2021). Distinctive traits for drought and salt stress tolerance in melon (*Cucumis melo* L.). *Frontiers in Plant Science 12*.

Clayberg, C.D. (1992). Interaction and linkage test of flesh colour genes in *Cucumis melo* L. *Cucurbit Genetics Cooperative 15*, 53.

Cohen, S., Itkin, M., Yeselson, Y., Tzuri, G., Portnoy, V., Harel-Baja, R., Lev, S., Sa'ar, U., Davidovitz-Rikanati, R., Baranes, N., *et al.* (2014). The PH gene determines fruit acidity and contributes to the evolution of sweet melons. *Nature Communications 5*, 4026.

Cohen, S., Tzuri, G., Harel-Beja, R., Itkin, M., Portnoy, V., Sa'ar, U., Lev, S., Yeselson, L., Petrikov, M., Rogachev, I., *et al.* (2012). Co-mapping studies of QTLs for fruit acidity and candidate genes of organic acid metabolism and proton transport in sweet melon (*Cucumis melo* L.). *Theoretical and Applied Genetics 125*, 343–353.

Dai, N., Cohen, S., Portnoy, V., Tzuri, G., Harel-Beja, R., Pompan-Lotan, M., Carmi, N., Zhang, G., Diber, A., Pollock, S., *et al.* (2011). Metabolism of soluble sugars in developing melon fruit: A global transcriptional view of the metabolic transition to sucrose accumulation. *Plant Molecular Biology 76*, 1–18.

Dai, X., Wang, Z., Bao, Y., Jia, C., Bai, F., Hasi, A., and Che, G. (2025). Identification and functional characterization of the C2H2 ZFP transcription factor *CmSUP7* in regulating melon plant growth and fruit development. *Plant Physiology and Biochemistry 220*, 190523.

Deleu, W., Esteras, C., Roig, C., González-To, M., Fernández-Silva, I., Gonzalez-Ibeas, D., Blanca, J., Aranda, M.A., Arús, P., Nuez, F., *et al.* (2009). A set of EST-SNPs for map saturation and cultivar identification in melon. *BMC Plant Biology 9*, 90.

Deol, J.K., Sharma, S.P., Rani, R., Kaila, A., Chhuneja, P., and Sarao, N.K. (2022). Inheritance analysis and identification of SSR markers associated with fusarium wilt resistance in melon. *The Journal of Horticultural Science and Biotechnology 97*, 66–74.

Diaz, A., Fergany, M., Formisano, G., Ziarsolo, P., Blanca, J., Fei, Z., Staub, J.E., Zalapa, J.E., Cuevas, H.E., and Dace, G.A. (2011). A consensus linkage map for molecular markers and Quantitative Trait Loci associated with economically important traits in melon (*Cucumis melo* L.). *BMC Plant Biology 11*.

Diaz, A., Forment, J., Argyris, J.M., Fukino, N., Tzuri, G., Harel-Beja, R., Katzir, N., Garcia-Mas, J., and Monforte, A.J. (2015). Anchoring the consensus ICuGI genetic map to the melon (*Cucumis melo* L.) genome. *Molecular Breeding 35*, 188.

Diaz-Pendon, J.A., Truniger, V., Nieto, C., Garcia-Mas, J., Bendahmane, A., and Aranda, M.A. (2004). Advances in understanding recessive resistance to plant viruses. *Molecular Plant Pathology 5*, 223–233.

Dos-Santos, N., Bueso, M.C., Díaz, A., Moreno, E., Garcia-Mas, J., Monforte, A.J., and Fernández-Trujillo, J.P. (2023). Thorough characterization of *ETHQB3.5*, a QTL involved in melon fruit climacteric behavior and aroma volatile composition. *Foods 12* (2), 376.

Ekbic, E., Fidan, H., Yildiz, M., and Abak, K. (2010). Screening of Turkish melon accessions for resistance to ZYMV, WMV and CMV. *Notulae Scientia Biologicae 2*, 55–57.

Feder, A., Burger, J., Gao, S., Lewinsohn, E., Katzir, N., Schaffer, A.A., Meir, A., Davidovich-Rikanati, R., Portnoy, V., Gal-On, A., *et al.* (2015). A Kelch domain-containing F-box coding gene negatively regulates flavonoid accumulation in *Cucumis melo* L. *Plant Physiology*, *16* (3), 1714–1726.

Frantz, J.D., and Jahn, M.M. (2003). Five independent loci each control monogenic resistance to gummy stem blight in melon (*Cucumis melo* L.). *Theoretical and Applied Genetics 108*, 1033–1038.

Fu, A., Zheng, Y., Guo, J., Grierson, D., Zhao, X., Wen, C., Liu, Y., Li, J., Zhang, X., Yu, Y., *et al.* (2023). Telomere-to-telomere genome assembly of bitter melon (*Momordica charantia* L. var. *abbreviata Ser.*) reveals fruit development, composition and ripening genetic characteristics. *Horticulture Research 10*, uhac228.

Galpaz, N., Gonda, I., Shem-Tov, D., Barad, O., Tzuri, G., Lev, S., Fei, Z., Xu, Y., Mao, L., Jiao, C., *et al.* (2018). Deciphering genetic factors that determine melon fruit-quality traits using RNA-Seq-based high-resolution QTL and eQTL mapping. *The Plant Journal 94*, 169–191.

Garcia-Mas, J., Benjak, A., Sanseverino, W., Bourgeois, M., Mir, G., González, V.M., Hénaff, E., Câmara, F., Cozzuto, L., Lowy, E., *et al.* (2012). The genome of melon (*Cucumis melo* L.). *Proceedings of the National Academy of Sciences 109*, 11872–11877.

Gonzalo, M.J., Oliver, M., Garcia-Mas, J., Monfort, A., Dolcet-Sanjuan, R., Katzir, N., Arús, P., and Monforte, A.J. (2005). Simple-sequence repeat markers used in merging linkage maps of melon (*Cucumis melo* L.). *Theoretical and Applied Genetics 110*, 802–811.

Guiu-Aragonés, C., Monforte, A.J., Saladié, M., Corrêa, R.X., Garcia-Mas, J., and Martín-Hernández, A.M. (2014). The complex resistance to cucumber mosaic cucumovirus (CMV) in the melon accession PI161375 is governed by one gene and at least two quantitative trait loci. *Molecular Breeding 34*, 351–362.

Harel-Beja, R., Tzuri, G., Portnoy, V., Lotan-Pompan, M., Lev, S., Cohen, S., Dai, N., Yeselson, L., Meir, A., Libhaber, S.E., *et al.* (2010). A genetic map of melon highly enriched with fruit quality QTLs and EST markers, including sugar and carotenoid metabolism genes. *Theoretical and Applied Genetics 121*, 511–533.

Hassan, M.Z., Rahim, M.A., Natarajan, S., Robin, A.H.K., Kim, H.-T., Park, J.-I., and Nou, I.-S. (2018). Gummy stem blight resistance in melon: Inheritance pattern and development of molecular markers. *International Journal of Molecular Sciences 19* (10), 2914.

Hu, J., Yao, J., Lu, J., Liu, W., Zhao, Z., Li, Y., Jiang, L., and Zha, L. (2024). The complete chloroplast genome sequences of nine melon varieties (*Cucumis melo* L.): Lights into comparative analysis and phylogenetic relationships. *Frontiers in Genetics 15*.

Hu, Z., Deng, G., Mou, H., Xu, Y., Chen, L., Yang, J., and Zhang, M. (2018). A re-sequencing-based ultra-dense genetic map reveals a gummy stem blight resistance-associated gene in *Cucumis melo*. *DNA Research 25*, 1–10.

Huang, H.-Y., Zhang, S., Choucha, F.A., Verdenaud, M., Tan, F.-Q., Pichot, C., Parsa, H.S., Slavkovic, F., Chen, Q., Troadec, C., *et al.* (2024). Harbinger transposon insertion in ethylene signaling gene leads to emergence of new sexual forms in cucurbits. *Nature Communications 15*, 4877.

Huang, L.T., Li-Na, Z., Li-Wei, G., Anne-Aliénor, V., Hervé, S., and Yi-Dong, Z. (2018). Constitutive expression of *CmSKOR*, an outward K+ channel gene from melon, in *Arabidopsis thaliana* involved in saline tolerance. *Plant Science 274*, 492–502.

Joobeur, T., King, J.J., Nolin, S.J., Thomas, C.E., and Dean, R.A. (2004). The fusarium wilt resistance locus Fom-2 of melon contains a single resistance gene with complex features. *The Plant Journal 39*, 283–297.

Kenigsbuch, D., and Cohen, Y. (1989). Independent inheritance of resistance to race 1 and race 2 of *Sphaerotheca fuliginea* in muskmelon. *Plant Disease 73*, 206–208.

Kishor, D.S., Noh, Y., Song, W.-H., Lee, G.P., Park, Y., Jung, J.-K., Shim, E.-J., Sim, S.-C., and Chung, S.-M. (2021). SNP marker assay and candidate gene identification for sex expression via genotyping-by-sequencing-based genome-wide associations (GWAS) analyses in Oriental melon (*Cucumis melo* L.var.*makuwa*). *Scientia Horticulturae 276.*

Latrasse, D., Rodriguez-Granados, N.Y., Veluchamy, A., Mariappan, K.G., Bevilacqua, C., Crapart, N., Camps, C., Sommard, V., Raynaud, C., Dogimont, C., *et al.* (2017). The quest for epigenetic regulation underlying unisexual flower development in *Cucumis melo*. *Epigenetics & Chromatin 10.*

Li, B., Zhao, Y., Zhu, Q., Zhang, Z., Fan, C., Amanullah, S., Gao, P., and Luan, F. (2017). Mapping of powdery mildew resistance genes in melon (*Cucumis melo* L.) by bulked segregant analysis. *Scientia Horticulturae 220*, 160–167.

Li, G., Tang, L., He, Y., Xu, Y., Bendahmane, A., Garcia-Mas, J., Lin, T., and Zhao, G. (2023). The haplotype-resolved T2T reference genome highlights structural variation underlying agronomic traits of melon. *Horticulture Research 10* (10), uhad182.

Li, L., Zhang, X., Ding, F., Hou, J., Wang, J., Luo, R., Mao, W., Li, X., Zhu, H., Yang, L., *et al.* (2024). Genome-wide identification of the melon (*Cucumis melo* L.) response regulator gene family and functional analysis of *CmRR6* and *CmPRR3* in response to cold stress. *Journal of Plant Physiology 292*, 154160.

Li, S., Wang, H., Li, Y., Jing, F., Xu, Y., Deng, S., Wang, N., Zhang, Z., and Chai, S. (2025). Mapping and functional characterization of the *golden fruit 1* (*gf1*) in melon (*Cucumis melo* L.). *Theoretical and Applied Genetics 138* (3), 59.

Lian, Q., Fu, Q., Xu, Y., Hu, Z., Zheng, J., Zhang, A., He, Y., Wang, C., Xu, C., Chen, B., *et al.* (2021). QTLs and candidate genes analyses for fruit size under domestication and differentiation in melon (*Cucumis melo* L.) based on high resolution maps. *BMC Plant Biology 21.*

Ling, J., Xie, X., Gu, X., Zhao, J., Ping, X., Li, Y., Yang, Y., Mao, Z., and Xie, B. (2021). High-quality chromosome-level genomes of Cucumis metuliferus and Cucumis melo provide insight into Cucumis genome evolution. *The Plant Journal 107*, 136–148.

Lippman, Z.B., and Zamir, D. (2007). Heterosis: Revisiting the magic. *Trends in Genetics 23*, 60–66.

Liu, B., Santo Domingo, M., Mayobre, C., Martín-Hernández, A.M., Pujol, M., and Garcia-Mas, J. (2022). Knock-Out of *CmNAC-NOR* affects melon climacteric fruit ripening. *Frontiers in Plant Science 13.*

Liu, L., Sun, T., Liu, X., Guo, Y., Huang, X., Gao, P., and Wang, X. (2019). Genetic analysis and mapping of a striped rind gene (*st3*) in melon (*Cucumis melo* L.). *Euphytica 215.*

Liu, S., Gao, P., Zhu, Q., Zhu, Z., Liu, H., Wang, X., Weng, Y., Gao, M., and Luan, F. (2020). Resequencing of 297 melon accessions reveals the genomic history of improvement and loci related to fruit traits in melon. *Plant Biotechnology Journal 18*, 2545–2558.

Liu, W., Jin, Y., Li, M., Dong, L., Guo, D., Lu, C., and Qi, H. (2018). Analysis of *CmCADs* and three lignifying enzymes in oriental melon ('CaiHong7') seedlings in response to three abiotic stresses. *Scientia Horticulturae 237*, 257–268.

Liu, W., Lv, Y., Zhang, L., Jiang, Y., Liu, S., Wang, Z., Zhang, J., and He, M. (2024). The gene *CmPYL6* strongly contributes to cold tolerance in oriental melon. *Plant Biology 26*, 1033–1046.

Lyu, X., Xia, Y., Wang, C., Zhang, K., Deng, G., Shen, Q., Gao, W., Zhang, M., Liao, N., Ling, J., *et al.* (2023). Pan-genome analysis sheds light on structural variation-based dissection of agronomic traits in melon crops. *Plant Physiology 193*, 1330–1348.

Marone, D., Russo, M., Laidò, G., De Leonardis, A., and Mastrangelo, A. (2013). Plant Nucleotide Binding Site–Leucine-Rich Repeat (NBS-LRR) Genes: Active Guardians in Host Defense Responses. *International Journal of Molecular Sciences 14*, 7302–7326.

Martin, A., Troadec, C., Boualem, A., Rajab, M., Fernandez, R., Morin, H., Pitrat, M., Dogimont, C., and Bendahmane, A. (2009). A transposon-induced epigenetic change leads to sex determination in melon. *Nature 461*, 1135–1138.

McCreight, J.D., Coffey, M.D., Turini, T.A., and Matheron, M.E. (2005). Field evidence for a new race of powdery mildew on melon. *Hortscience 40*, 888.

Mizuno, S., Masuda, C., Otsuka, A., Kishimoto, N., Kameyama, C., Kamiyoshihara, Y., and Mitsuzawa, H. (2024). Interaction between plant-specific transcription factors TCP and YABBY expressed in the tendrils of the melon *Cucumis melo*. *Scientific Reports 14*.

Mo, C., Wang, H., Wei, M., Zeng, Q., Zhang, X., Fei, Z., Zhang, Y., and Kong, Q. (2024). Complete genome assembly provides a high-quality skeleton for pan-NLRome construction in melon. *The Plant Journal 118*, 2249–2268.

Moing, A., Allwood, J.W., Aharoni, A., Baker, J., Beale, M.H., Ben-Dor, S., Biais, B., Brigante, F., Burger, Y., Deborde, C., *et al.* (2020). Comparative metabolomics and molecular phylogenetics of melon (*Cucumis melo, Cucurbitaceae*) biodiversity. *Metabolites 10* (3): 121.

Nashiki, A., Matsuo, H., Takano, K., Fitriyah, F., Isobe, S., Shirasawa, K., and Yoshioka, Y. (2023). Identification of novel sex determination loci in Japanese weedy melon. *Theoretical and Applied Genetics 136* (6), 136.

Nimmakayala, P., Tomason, Y.R., Abburi, V.L., Alvarado, A., Saminathan, T., Vajja, V.G., Salazar, G., Panicker, G.K., Levi, A., Wechter, W.P., *et al.* (2016). Genome-wide differentiation of various melon horticultural groups for use in GWAS for fruit firmness and construction of a high resolution genetic map. *Frontiers in Plant Science 7*.

Ning, X., Wang, X., Gao, X., Zhang, Z., Zhang, L., Yan, W., and Li, G. (2013). Inheritances and location of powdery mildew resistance gene in melon Edisto47. *Euphytica 195*, 345–353.

Nonaka, S., Ito, M., and Ezura, H. (2023). Targeted modification of *CmACO1* by CRISPR/Cas9 extends the shelf-life of Cucumis melo var. reticulatus melon. *Frontiers in Genome Editing 5*.

Nuñez-Palenius, H.G., Gomez-Lim, M., Ochoa-Alejo, N., Grumet, R., Lester, G., and Cantliffe, D.J. (2008). Melon fruits: Genetic diversity, physiology, and biotechnology features. *Critical Reviews in Biotechnology 28*, 13–55.

Oren, E., Dafna, A., Tzuri, G., Halperin, I., Isaacson, T., Elkabetz, M., Meir, A., Saar, U., Ohali, S., La, T., *et al.* (2022). Pan-genome and multi-parental framework for high-resolution trait dissection in melon (*Cucumis melo*). *The Plant Journal 112*, 1525–1542.

Oren, E., Tzuri, G., Vexler, L., Dafna, A., Meir, A., Faigenboim, A., Kenigswald, M., Portnoy, V., Schaffer, A.A., Levi, A., *et al.* (2019). The multi-allelic *APRR2* gene is associated with fruit pigment accumulation in melon and watermelon. *Journal of Experimental Botany 70*, 3781–3794.

Oumouloud, A., and Álvarez, J.M. (2016). Breeding and genetics of resistance to Fusarium wilt in melon. In *Advances in Plant Breeding Strategies: Agronomic, Abiotic and Biotic Stress Traits*, 601–626. Springer, Cham.

Oumouloud, A., Arnedo-Andrés, M.S., González-Torres, R., and Álvarez, J.M. (2010). Inheritance of resistance to *Fusarium oxysporum* f. sp. *melonis* races 0 and 2 in melon accession Tortuga. *Euphytica 176*, 183–189.

Pascual, L., Yan, J., Pujol, M., Monforte, A.J., Picó, B., and Martín-Hernández, A.M. (2019). *CmVPS41* is a general gatekeeper for resistance to cucumber *Mosaic Virus* phloem entry in melon. *Frontiers in Plant Science 10*.

Perchepied, L., Bardin, M., Dogimont, C., and Pitrat, M. (2005). Relationship between loci conferring downy mildew and powdery mildew resistance in melon assessed by quantitative trait loci mapping. *Phytopathology 95*, 556–565.

Pereira, L., Ruggieri, V., Pérez, S., Alexiou, K.G., Fernández, M., Jahrmann, T., Pujol, M., and Garcia-Mas, J. (2018). QTL mapping of melon fruit quality traits using a high-density GBS-based genetic map. *BMC Plant Biology 18*, 324.

Pichot, C., Djari, A., Tran, J., Verdenaud, M., Marande, W., Huneau, C., Gautier, V., Latrasse, D., Arribat, S., Sommard, V., *et al.* (2022). Cantaloupe melon genome reveals 3D chromatin features and structural relationship with the ancestral cucurbitaceae karyotype. *iScience 25* (1), 103696.

Prasad, K., and Norton, J.D. (1967). Inheritance of resistance to *Mycosphaerella citrullina* in muskmelon. *Horticultural Science 91*, 396–400.

Rashid, D., Devani, R.S., Rodriguez-Granados, N.Y., Abou-Choucha, F., Troadec, C., Morin, H., Tan, F.-Q., Marcel, F., Huang, H.-Y., Hanique, M., *et al.* (2023). Ethylene produced in carpel primordia controls *CmHB40* expression to inhibit stamen development. *Nature Plants 9*, 1675–1687.

Ren, K., Kong, W., Tang, T., and Cheng, H. (2023). Identification of *CmACL* genes in melon and analysis of their potential functions in fruit sugar and acid accumulation. *Frontiers in Plant Science 14*.

Ren, T., Shi, X., Zhou, S., Fan, K., Zhang, R., Nie, L., and Zhao, W. (2025). Transcriptome profiling reveals the mechanism of fruit navel development in melon (*Cucumis melo* L.). *BMC Plant Biology 25*, 420.

Rodriguez-Hernandez, A.M., Gosalvez, B., Sempere, R.N., Burgos, L., Aranda, M.A., and Truniger, V. (2012). melon RNA interference (RNAi) lines silenced for *Cm-eIF4E* show broad virus resistance. *Molecular Plant Pathology 13*, 755–763.

Ruggieri, V., Alexiou, K.G., Morata, J., Argyris, J., Pujol, M., Yano, R., Nonaka, S., Ezura, H., Latrasse, D., Boualem, A., *et al.* (2018). An improved assembly and annotation of the melon (*Cucumis melo* L.) reference genome. *Scientific Reports 8*.

Sanseverino, W., Hénaff, E., Vives, C., Pinosio, S., Burgos-Paz, W., Morgante, M., Ramos-Onsins, S.E., Garcia-Mas, J., and Casacuberta, J.M. (2015). Transposon insertions, structural variations, and SNPs contribute to the evolution of the melon genome. *Molecular Biology and Evolution 32*, 2760–2774.

Saxena, R.K., Edwards, D., and Varshney, R.K. (2014). Structural variations in plant genomes. *Briefings in Functional Genomics 13*, 296–307.

Schemberger, M.O., Stroka, M.A., Reis, L., de Souza Los, K.K., de Araujo, G.A.T., Sfeir, M.Z.T., Galvão, C.W., Etto, R.M., Baptistão, A.R.G., and Ayub, R.A. (2020). Transcriptome profiling of non-climacteric 'yellow' melon during ripening: Insights on sugar metabolism. *BMC Genomics 21*, 262.

Shen, J., Xu, X., Zhang, Y., Niu, X., and Shou, W. (2021). Genetic mapping and identification of the candidate genes for mottled rind in *Cucumis melo* L. *Frontiers in Plant Science 12*.

Shin, A.-Y., Koo, N., Kim, S., Sim, Y.M., Choi, D., Kim, Y.-M., and Kwon, S.-Y. (2019). Draft genome sequences of two oriental melons, *Cucumis melo* L. var. *makuwa*. *Scientific Data 6*.

Sun, Y., Wang, J., Li, Y., Jiang, B., Wang, X., Xu, W.-H., Wang, Y.-Q., Zhang, P.-T., Zhang, Y.-J., and Kong, X.-D. (2022). Pan-genome analysis reveals the abundant gene presence/absence variations among different varieties of melon and their influence on traits. *Frontiers in Plant Science 13*.

Tang, L., He, Y., Liu, B., Liu, M., Xu, Y., Zhang, J., Kong, W., An, L., Hu, K., Garcia-Mas, J., Liu, B., and Zhao, G. (2025). *CmFUL1* was potentially involved in fruit elongation in melon. *Horticulture Research, 12* (8), uhaf138.

Tang, L., He, Y., Liu, B., Xu, Y., and Zhao, G. (2023). Genome-wide identification and characterization analysis of WUSCHEL-related homeobox family in melon (*Cucumis melo* L.). *International Journal of Molecular Sciences 24* (15), 12326.

Tang, M., Bie, Z.-l., Wu, M.-z., Yi, H.-p., and Feng, J.-x. (2010). Changes in organic acids and acid metabolism enzymes in melon fruit during development. *Scientia Horticulturae 123*, 360–365.

Tezuka, T., Waki, K., Yashiro, K., Kuzuya, M., Ishikawa, T., Takatsu, Y., and Miyagi, M. (2009). Construction of a linkage map and identification of DNA markers linked to *Fom-1*, a gene conferring resistance to *Fusarium oxysporum* f.sp. *melonis* race 2 in melon. *Euphytica 168*, 177–188.

Thakur, H., Sharma, S., and Thakur, M. (2019). Biotechnology: Recent trends in muskmelon (*Cucumis melo* L.) research: An overview. *Journal of Horticultural Science and Biotechnology 94*, 533–547.

Tian, D., Traw, M., Chen, J., Kreitman, M., and Bergelson, J. (2003). Fitness costs of *R*-gene-mediated resistance in *Arabidopsis thaliana*. *Nature 423*, 74–77.

Tieman, D., Zhu, G., Resende Jr., M.F.R., Lin, T., Nguyen, C., Bies, D., Rambla, J.L., Beltran, K.S.O., Taylor, M., Zhang, B., *et al.* (2017). A chemical genetic roadmap to improved tomato flavor. *Science 335*, 391–394.

Toporek, S.M., Branham, S.E., Keinath, A.P., and Wechter, W.P. (2023). QTL mapping of resistance to Pseudoperonospora cubensis clade 2, mating type A1, in Cucumis melo and dual-clade marker development. *Theoretical and Applied Genetics 136* (4), 91.

Truniger, V., and Aranda, M.A. (2009). Recessive resistance to plant viruses. *Advances in Virus Research 75*, 119–231.

Tzuri, G., Zhou, X., Chayut, N., Yuan, H., Portnoy, V., Meir, A., Sa'ar, U., Baumkoler, F., Mazourek, M., Lewinsohn, E., *et al.* (2015). A 'golden' SNP in *CmOr* governs the fruit flesh color of melon (*Cucumis melo*). *The Plant Journal 82*, 267–279.

Virtuoso, M.C.S., Valente, T.S., Silva, E.H.C., Braz, L.T., Panizzi, R.d.C., and Vargas, P.F. (2022). Implications of the inoculation method and environment in the selection of melon genotypes resistant to *Didymella bryoniae*. *Scientia Horticulturae 300*, 111066.

Wang, P., Xu, X., Zhao, G., He, Y., Hou, C., Kong, W., Zhang, J., Liu, S., Xu, Y., and Xu, Z. (2020). Genetic mapping and candidate gene analysis for melon resistance to *Phytophthora capsici*. *Scientific Reports 10*.

Wang, R., Tavano, E.C.d.R., Lammers, M., Martinelli, A.P., Angenent, G.C., and de Maagd, R.A. (2019). Re-evaluation of transcription factor function in tomato fruit development and ripening with CRISPR/Cas9-mutagenesis. *Scientific Reports 9*.

Wang, Y., Thomas, C., and Dean, R. (1997). A genetic map of melon (*Cucumis melo* L.) based on amplified fragment length polymorphism (AFLP) markers. *Theoretical and Applied Genetics 95*, 791–798.

Wei, M., Huang, Y., Mo, C., Wang, H., Zeng, Q., Yan.g, W., Chen, J., Zhang, X., and Kong, Q. (2023). Telomere-to-telomere genome assembly of melon (*Cucumis melo* L. var. *inodorus*) provides a high-quality reference for meta-QTL analysis of important traits. *Horticulture Research 10* (10), uhad189.

Wu, X., Wang, L., Xing, Q., Zhao, Y., and Qi, H. (2024). CmPIF8-CmERF27-*CmACS10*-mediated ethylene biosynthesis modulates red light-induced powdery mildew resistance in oriental melon. *Plant, Cell & Environment 47*, 4135–4150.

Xia, Y., Xu, M., Yang, Q., Zhang, Y., Shen, Q., Zhao, H., Zhang, K., Meng, Y., Batool, A., Bo, Y., *et al.* (2025). *CmPPR4* gene controls drought resilience in melon ecotypes. *Plant Biotechnology Journal 23*, 1881–1891.

Xing, Q., Liao, J., Cao, S., Li, M., Lv, T., and Qi, H. (2020). *CmLOX10* positively regulates drought tolerance through jasmonic acid -mediated stomatal closure in oriental melon (*Cucumis melo* var. *makuwa Makino*). *Scientific Reports 10*.

Xu, D., Han, Y., Zhang, Y., Khan, A., Dong, L., Shao, L., Liang, A., Liu, T., and Qi, H. (2025). *CmTGA8-CmAPX1/CmGSTU25* regulatory model involved in trehalose induced cold tolerance in oriental melon seedlings. *Plant Physiology and Biochemistry 220*, 109432.

Yan, H., Wang, K., Wang, M., Feng, L., Zhang, H., and Wei, X. (2023). QTL mapping and genome-wide association study reveal genetic loci and candidate genes related to soluble solids content in Melon. *Current Issues in Molecular Biology 45*, 7110–7129.

Yang, J., Deng, G., Lian, J., Garraway, J., Niu, Y., Hu, Z., Yu, J., and Zhang, M. (2020). The chromosome-scale genome of melon dissects genetic architecture of important agronomic traits. *iScience 23* (8), 101422.

Yano, R., Ariizumi, T., Nonaka, S., Kawazu, Y., Zhong, S., Mueller, L., Giovannoni, J.J., Rose, J.K.C., and Ezura, H. (2020). Comparative genomics of muskmelon reveals a potential role for retrotransposons in the modification of gene expression. *Communications Biology 3*.

Zhang, C., Ren, Y., Guo, S., Zhang, H., Gong, G., Du, Y., and Xu, Y. (2012). Application of comparative genomics in developing markers tightly linked to the Pm-2F gene for powdery mildew resistance in melon (*Cucumis melo* L.). *Euphytica 190*, 157–168.

Zhang, H., Li, X., Yu, H., Zhang, Y., Li, M., Wang, H., Wang, D., Wang, H., Fu, Q., Liu, M., *et al.* (2019). A high-quality melon genome assembly provides insights into genetic basis of fruit trait improvement. *iScience 22*, 16–27.

Zhang, T., Ding, Z., Liu, J., Qiu, B., and Gao, P. (2020). QTL mapping of pericarp and fruit-related traits in melon (*Cucumis melo* L.) using SNP-derived CAPS markers. *Scientia Horticulturae 265*, 109243.

Zhang, T., Xu, N., Amanullah, S., and Gao, P. (2023a). Genome-wide identification, evolution, and expression analysis of *MLO* gene family in melon (*Cucumis melo* L.). *Frontiers in Plant Science 14*.

Zhang, X., Ling, Y., Yang, W., Wei, M., Wang, Z., Li, M., Yang, Y., Liu, B., Yi, H., Guo, Y.-D., *et al.* (2023b). Fine mapping of a novel QTL *DM9.1* conferring downy mildew resistance in melon. *Frontiers in Plant Science 14*.

Zhao, G., Lian, Q., Zhang, Z., Fu, Q., He, Y., Ma, S., Ruggieri, V., Monforte, A.J., Wang, P., Julca, I., *et al.* (2019). A comprehensive genome variation map of melon identifies multiple domestication events and loci influencing agronomic traits. *Nature Genetics 51*, 1607–1615.

Zhao, H., Zhang, T., Meng, X., Song, J., Zhang, C., and Gao, P. (2023). Genetic mapping and QTL analysis of fruit traits in melon (*Cucumis melo* L.). *Current Issues in Molecular Biology 45*, 3419–3433.

Zheng, L., Dai, H., Mu, Y., Li, J., Cheng, Y., and Han, J. (2025). Genome-wide identification and expression analysis of *C3H* gene family in melon. *Frontiers in Plant Science 16*.

Zink, F., and Gubler, W.D. (1985). Inheritance of resistance in muskmelon to Fusarium wilt. *Journal of the American Society for Horticultural Science 110*, 600–604.

4 Genomics of Sex Expression – An Important Factor in Melon Breeding

Akito Nashiki

4.1 INTRODUCTION

Flower development is crucial for the survival strategy in plants, and it also plays a critical role in yield, fruit quality, and seed production. In angiosperms, more than 90% of species have hermaphrodite (bisexual flowers), which have pistils and stamens simultaneously, whereas very few plant species are dioecious (male and female plants) or monoecious (male and female flowers on the same plant; Ainsworth 2000; Barrett 2002; Renner 2014). Thus, sexual systems have high diversity and have long been a fascinating research topic for biologists. Recently, the molecular and genetic mechanisms of sex determination have been studied well in some plants, including persimmon (*Diospyros* spp.; Akagi et al. 2014, 2020; Masuda et al. 2019), kiwifruit (*Actinida* spp.; Akagi et al. 2018, 2019), and garden asparagus (*Asparagus officinalis* L.; Harkess et al. 2017, 2020; Murase et al. 2017; Tsugama et al. 2017).

In Cucurbitaceae, there are many economically important crops, such as cucumber (*Cucumis sativus*), melon (*Cucumis melo*), watermelon (*Citrullus lanatus*), zucchini (*Cucurbita pepo*), and pumpkins (*Cucurbita moschata*; Bhowmick and Jha 2015). Melon is one of the most economically important horticultural crops and has long been cultivated and grown worldwide. It has extreme diversity in terms of fruit size, fruit shape, sweetness, aroma, and sex expression (Monforte et al. 2014; Girek et al. 2013; Revanasidda and Belavadi 2019). Of sex expression, Cucurbitaceae has three flower types (bisexual, female, male), and the distribution of the three flower types in the same plant produces various sexual morphologies: monoecious, andromonoecious, hermaphrodite, gynoecious, androecious, and trimonoecious. So far, many researchers have cloned the genes related to sex determination and revealed the regulation of plant hormones (Latrasse et al. 2017; Boualem et al. 2008, 2015, 2022; Rodriguez-Granados et al. 2022). In Cucurbitaceae, male and female flowers differentiate from initially bisexual flowers by a selective developmental arrest of either the development of the stamen or the carpel primordia (Kater et al. 2001; Hao et al. 2003; Bai et al. 2004). All the sex determination genes that have been cloned to date are involved indirectly or directly in the ethylene signaling pathway, and ethylene is

 DOI: 10.1201/9781003715757-4

known to be key in sex determination in cucurbit crops (Byers et al. 1972; Boualem et al. 2008, 2015, 2022; Chen et al. 2016; Martin et al. 2009; Zhang et al. 2022; Rashid et al. 2023). Local expression of sex determination genes in the carpels and stamens results in the final flower sex. However, there are still limited reports of the mechanism regarding how different sexual flowers emerge locally on the same plant in melon (Latrasse et al. 2017; Rodriguez-Granados et al. 2022). An elucidation of the network of hormones and genes related to sex determination in cucurbits will not only provide important knowledge for evolutionary research on plant sexuality but will have critical applications in plant breeding and crop production. In cucurbits, sex determination is crucial for controlling yield, enhancing fruit quality, and organizing breeding plans. In this chapter, we review sex determination studies in melons and their potential for use in breeding.

4.2 FACTORS REGULATING SEX EXPRESSION IN MELON

Until recently, in melon, a sex determination model consisting of three sex determination genes, including *CmACS7* (Boualem et al. 2008), *CmACS11* (Boualem et al. 2015), and *CmWIP1* (Martin et al. 2009), has been proposed (Figure 4.1). All these genes are responsible for the ethylene biosynthesis pathway, and *CmACS7* and *CmACS11* are among the genes encoding for a 1-aminocyclopropane-1-carboxylic acid synthase. *CmACS7* regulates stamen development, and normal enzymatic activity (*M* allele) of the CmACS7 protein inhibits the development of stamen primordia, resulting in female flowers, whereas a reduction of the enzymatic activity (*m* allele) results in hermaphrodite flowers (Boualem et al. 2008). *CmWIP1* encodes a C2H2 zinc finger–type transcription factor that regulates pistil development, and the presence of the transposon *Gyno-hAT* in the promoter region of *CmWIP1* (*g* allele) methylates *CmWIP1* and promotes pistil development. In the absence of *Gyno-hAT* (*G* allele), *CmWIP1* inhibits the development of pistil primordia and suppresses the expression of the *M* allele of *CmACS7*, which regulates the development of stamen primordia (Martin et al. 2009). *CmACS11* (*a* allele) is expressed only in lateral branches and promotes pistil development by suppressing the expression of *CmWIP1* (*G* allele), while it is not expressed in main branches and promotes the expression of *CmWIP1* (*G* allele), which suppresses pistil development and the expression of *CmACS7*, resulting in a male flower. In addition, a loss-of-function Ethyl methanesulfonate (EMS)-induced mutant of *CmACS11* (*a* allele) does not express the gene in main and lateral branches, allowing expression of *CmWIP1* (*G* allele) and promoting stamen development, resulting in male flowers at all nodes (Boualem et al. 2015). Thus, *CmACS11* is responsible for the coexistence of different floral characters on the same plant (dioecism) by producing bisexual or female flowers on the lateral branches and male flowers on the main branches. In this sex determination model by three genes, the allelic combinations of them have been proposed to produce a variety of sex expressions, including andromonoecious (*A-G-mm*), monoecious (*A-G-M-*), gynoecious (*-ggM-*), hermaphrodite (*-ggmm*), and androecious (*aaGG-*).

From 2022, *CmCRC* (Zhang et al. 2022) on chromosome 10, *CmCPR5* (Wang et al. 2022), and *CmHB40* (Rashid et al. 2023) on chromosome 8 have been newly

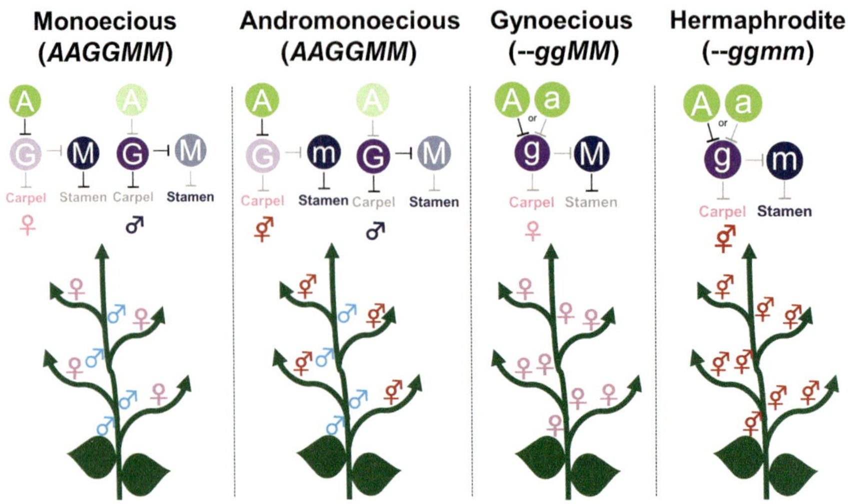

FIGURE 4.1 Sex expression model in melon by 2022. A/a: *CmACS11*, G/g: *CmWIP1*, M/m: *CmACS7*.

identified as sex determination genes. The expression of *CmCRC* is regulated by the binding of CmWIP1 protein to the promoter region of *CmCRC*, and the simultaneous expression of *CmACS7* (*M* allele) leads to male flowers emerging. When *CmACS11* suppresses *CmWIP1* expression, *CmCRC* expresses in the carpel primordium and promotes carpel development, while *CmACS11* suppresses the expression of *CmACS7* (*M* allele) and inhibits stamen development, resulting in female flowers (Zhang et al. 2022). *CmCPR5*, a transcript from the dominant allele (*B* allele), interacts with CmETR1, an ethylene response factor, in the stamen primordium to promote ethylene signaling, resulting in a female flower. However, in the case of the recessive allele (*b* allele), the ethylene response is disrupted by the inability to regulate *CmETR1*, which is proposed to promote the development of stamens by inhibiting ethylene production by *CmACS7* (*M* allele), which inhibits the development of stamen primordia, resulting in bisexual flowers (Wang et al. 2022). Rashid et al. (2023) proposed that a mutation in *CmHB40* (*Cmhb40*^{W113}), which is downstream of the ethylene signaling pathway produced by *CmACS7* (*M* allele), promotes stamen development. In addition, genetic resources that exhibit andromonoecious have been found in Africa (Kenya, Sudan, and Zimbabwe) despite the presence of this mutant allele of *Cmhb40*^{W113} and the *M* allele of *CmACS7*. Thus, all the previously reported sex determination genes are directly or indirectly responsible for the ethylene signaling pathway, and the allelic combination of each gene during each developmental stage of the carpel or stamen primordium affects the biosynthesis of the final product, ethylene, and finally determines the sex. However, there have been no reports of developing plants with all allelic combinations for the previously reported genes, and the mechanism remains unknown. In addition, Nashiki et al. (2023) reported a locus that produces pistil-bearing flowers (female/bisexual) on the main branch, and the combination of the allele and previously reported sex determination genes changes sex expression. For the locus, it is not clear how it interacts with other genes or even whether it is involved in ethylene biosynthesis. To reveal the molecular mechanisms, more research is required to unravel organ programs, without disrupting flower patterning (Coen and Meyerowitz 1991).

In cucurbits, interestingly, the association of fruit shape and sex expression has long been reported (Wall 1967; Rosa 1928; Périn et al., 2002; Loy 2006; Abdelmohsin and Pitrat 2008; Sakata et al. 2013). Most of the cucumber and melon are female flowers, and their fruits usually show an elongated shape, while bisexual flowers bear round fruits. This association of a bisexual flower and a round fruit shape has been due to the pleiotropic effects of the monoecy gene, *CsACS2/CmACS7*, rather than due to its close linkage with a fruit shape gene (Abdelmohsin and Pitrat 2008), and the genetic and physiological mechanism of this association was not well understood. However, recently, Boualem et al. (2022) revealed *CmACS7*-mediated ethylene production in carpel primordia enhances cell expansion and represses cell division, leading to elongated fruit. Furthermore, Rashid et al. (2023) identified the presence of a novel gene, *CmHB40*, that disrupts the association between *CmACS7* and fruit shape. Thus, while there is a strong association between sex expression and fruit shape, the possibility of untangling the association has emerged, potentially paving the way for the application of sex determination genes for melon breeding.

4.3 APPLICATION OF SEX DETERMINATION GENES IN BREEDING

Sex expression influences fruit quality, yield, cultivation method, and breeding strategy. Monoecy is of agronomic interest both for breeding strategy and for the seed production due to three main reasons: (1) Emasculation is not required to produce hybrid seed, (2) the facility to produce hybrid seeds ensures confidentiality, and (3) fruits produced by monoecious hybrids develop a smaller bottom scar, resulting in a reduced risk of pathogen infections and better fruit quality (Périn et al. 2002; Noguera et al. 2005). Gynoecy is also a valuable trait, as the number of female flowers directly correlates with yield in cucurbits, such as cucumber and bitter gourd, enabling the harvest of numerous immature fruits and eliminating the waiting period for the lateral branches to elongate and facilitate fruit set on the main stem. In cucumber, monoecy of the main stem is a dominant trait (Li et al. 2020), whereas in bitter gourd, it is a recessive trait (Matsumura et al. 2014). Conversely, in melon, the trait of bearing a pistillate flower on the main stem is a recessive trait, determined by a loss-of-function mutation in *CmWIP1* (*g* allele; Martin et al. 2009) and a dominant trait, determined by Mendelian genetic analyses in *Opbf3.1* (Nashiki et al. 2023). The flowering time of the main stem is approximately one week earlier than that of the first node of the lateral branches emerging from the same node, allowing fruit set on the main stem to shorten the cultivation period. Additionally, when fruit is set on the main stem, all lateral branches can be pruned, reducing the cultivation area required per plant. This facilitates dense planting, thereby increasing yield per unit area. Based on these findings, cultivars with the trait of bearing pistillate flowers on the main stem are expected to contribute to labor-saving cultivation practices and increased fruit yield. In hybrid melon seed production, gynoecy ensures hybrid seed purity and allows cross-pollination by pollinator insects. Therefore, the breeding application of gynoecy lines has been proposed for a long time (Kenigsbuch and Cohen 1990; Loy et al. 1979). However, all previously developed gynoecious lines have been derived from the *CmWIP1* loss-of-function allele found in the genetic resource WI998 (Peterson et al. 2022; Cohen et al. 1993). These lines exhibit unstable gynoecious expression under high-temperature conditions, leading to the frequent development of staminate flowers, thereby limiting their practical applications (Kesh and Kaushik 2021). Consequently, there is a strong demand for the discovery of novel genes that contribute to the stabilization of gynoecy and the development of fully gynoecious lines.

4.4 CONCLUDING REMARKS

The remarkable sexual diversity observed in melon offers valuable perspectives on evolutionary and developmental biology while opening new possibilities for crop improvement. Currently, a core ethylene-based sex determination mechanism has been proposed in cucurbit crops, but the details of the direct regulators and molecular mechanisms remain poorly understood. Recently, new variants of melon have been found and new aspects of the mechanism have been discovered (Rashid et al. 2023; Nashiki et al. 2023; Wang et al. 2022). In the future, evaluating the phenotypic stability of sex determination genes under environmental factors will be crucial for their application in breeding programs. We hope that the suggestions proposed in this review will effectively help apply sex expression to cucurbit breeding programs.

REFERENCES

Abdelmohsin, E. M., Pitrat, M. 2008. "Pleiotropic Effect of Sex Expression on Fruit Shape in Melon." In Pitrat, M., ed. *9th EUCARPIA Meeting on Genetics and Breeding of Cucurbitaceae*. Avignon, France, 551–556.

Ainsworth, C. 2000. "Boys and Girls Come Out to Play: The Molecular Biology of Dioecious Plants." *Annals of Botany* 86 (2): 211–21. https://doi.org/10.1006/anbo.2000.1201.

Akagi, T., Henry, I. M., Ohtani, H., Morimoto, T., Beppu, K., Kataoka, I., Tao, R. 2018. "A Y-Encoded Suppressor of Feminization Arose Via Lineage-Specific Duplication of a Cytokinin Response Regulator in Kiwifruit." *Plant Cell* 30: 780–95. https://doi.org/10.1105/tpc.17.00787.

Akagi, T., Henry, I. M., Tao, R., Comai, L. 2014. "A Y-Chromosome-Encoded Small RNA Acts as a Sex Determinant in Persimmons." *Science* 346: 646–50. https://doi.org/10.1126/science.1257225.

Akagi, T., Pilkington, S. M., Varkonyi-Gasic, E., Henry, I. M., Sugano, S. S., Sonoda, M., Firl, A., McNeilage, M. A., Douglas, M. J., Wang, T., Rebstock, R., Voogd, C., Datson, P., Allan, A. C., Beppu, K., Kataoka, I., Tao, R. 2019. "Two Y-Chromosome-Encoded Genes Determine Sex in Kiwifruit." *Nature Plants* 5: 801–9. https://doi.org/10.1038/s41477-019-0489-6.

Akagi, T., Shirasawa, K., Nagasaki, H., Hirakawa, H., Tao, R., Comai, L., Henry, I. M. 2020. "The Persimmon Genome Reveals Clues to the Evolution of a Lineage-Specific Sex Determination System in Plants." *PLoS Genetics* 16: 1–26. https://doi.org/10.1371/journal.pgen.1008566.

Bai, S. L., Peng, Y. B., Cui, J. X., Gu, H. T., Xu, L. Y., Li, Y. Q., Xu, Z. H., Bai, S. N. 2004. "Developmental Analyses Reveal Early Arrests of the Spore-Bearing Parts of

Reproductive Organs in Unisexual Flowers of Cucumber (*Cucumis sativus* L.)." *Planta* 220: 230–40. https://doi.org/10.1007/s00425-004-1342-2.

Barrett, S. C. H. 2002. "The Evolution of Plant Sexual Diversity." *Nature Reviews. Genetics* 3 (4): 274–84. https://doi.org/10.1038/nrg776.

Bhowmick, B. K., Jha, S. 2015. "Dynamics of Sex Expression and Chromosome Diversity in Cucurbitaceae: A Story in the Making." *Journal of Genetics* 94: 793–808. https://doi.org/10.1007/s12041-015-0562-5.

Boualem, A., Berthet, S., Devani, R. S., Camps, C., Fleurier, S., Morin, H., Troadec, C., Giovinazzo, N., Sari, N., Dogimont, C., Bendahmane, A. 2022. "Ethylene Plays a Dual Role in Sex Determination and Fruit Shape in Cucurbits." *Current Biology* 32 (11): 2390–401.e4. https://doi.org/10.1016/j.cub.2022.04.031.

Boualem, Adnane, Fergany, M., Fernandez, R., Troadec, C., Martin, A., Morin, H., Sari, M. A., Collin, F., Flowers, J.-M., Pitrat, M., Purugganan, M. D., Dogimont, C., Bendahmane, A. 2008. "A Conserved Mutation in an Ethylene Biosynthesis Enzyme Leads to Andromonoecy in Melons." *Science* 321 (5890): 836–8. https://doi.org/10.1126/science.1159023.

Boualem, Adnane, Troadec, C., Camps, C., Lemhemdi, A., Morin, H., Sari, M.-A., Fraenkel-Zagouri, R., Kovalski, I., Dogimont, C., Perl-Treves, R., Bendahmane, A. 2015. "A Cucurbit Androecy Gene Reveals How Unisexual Flowers Develop and Dioecy Emerges." *Science* 350 (6261): 688–91. https://doi.org/10.1126/science.aac8370.

Byers, R. E., Baker, L. R., Sell, H. M., Herner, R. C., Dilley, D. R. 1972. "Ethylene: A Natural Regulator of Sex Expression of *Cucumis melo* L." *Proceedings of the National Academy of Sciences of the United States of America* 69 (3): 717–20. https://doi.org/10.1073/pnas.69.3.717.

Chen, Huiming, Sun, J., Li, S., Cui, Q., Zhang, H., Xin, F., Wang, H., Lin, T., Gao, S., Wang, S., Li, X., Wang, S., Zhang, Z., Xu, Z., Huang, S. 2016. "An ACC Oxidase Gene Essential for Cucumber Carpel Development." *Molecular Plant* 9 (9): 1315–27. https://doi.org/10.1016/j.molp.2016.06.018.

Coen, E. S., Meyerowitz, E. M. 1991. "The War of the Whorls: Genetic Interactions Controlling Flower Development." *Nature* 353 (6339): 31–7. https://doi.org/10.1038/353031a0.

Cohen, Y., Eyal, H., Cohen, A. 1993. " 'Gylan' – A Gynoecious Muskmelon." *HortScience* 28 (8): 855. https://doi.org/10.21273/HORTSCI.28.8.855.

Girek, Z., Prodanovic, S., Zdravkovic, J., Zivanovic, T., Ugrinovic, M., Zdravkovic, M. 2013. "The Effect of Growth Regulators on Sex Expression in Melon (*Cucumis melo* L.). Crop Breed." *Journal of Applied Biology* 13: 165–71.

Hao, Y.-J., Wang, D.-H., Peng, Y.-B., Bai, S. L., Xu, L.-Y., Li, Y.-Q., Xu, Z.-H., Bai, S.-N. 2003. "DNA Damage in the Early Primordial Anther Is Closely Correlated with Stamen Arrest in the Female Flower of Cucumber (*Cucumis sativus* L.)." *Planta* 217 (6): 888–95. https://doi.org/10.1007/s00425-003-1064-x.

Harkess, A., Huang, K., van der Hulst, R., Tissen, B., Caplan, J. L., Koppula, A., Batish, M., Meyers, B. C., Leebens-Mack, J. 2020. "Sex Determination by Two Y-Linked Genes in Garden Asparagus." *Plant Cell* 32 (6): 1790–6. https://doi.org/10.1105/tpc.19.00859.

Harkess, A., Zhou, J., Xu, C., Bowers, J. E., Van Der Hulst, R., Ayyampalayam, S., Mercati, F., Riccardi, P., McKain, M. R., Kakrana, A., Tang, H., Ray, J., Groenendijk, J., Arikit, S., Mathioni, S. M., Nakano, M., Shan, H., Telgmann-Rauber, A., Kanno, A., Yue, Z., Chen, Haixin, Li, W., Chen, Y., Xu, X., Zhang, Y., Luo, S., Chen, Helong, Gao, J., Mao, Z., Pires, J. C., Luo, M., Kudrna, D., Wing, R. A., Meyers, B. C., Yi, K., Kong, H., Lavrijsen, P., Sunseri, F., Falavigna, A., Ye, Y., Leebens-Mack, J. H., Chen, G. 2017. "The Asparagus Genome Sheds Light on the Origin and Evolution of a Young Y Chromosome." *Nature Communications* 8:1279. https://doi.org/10.1038/s41467-017-01064-8.

Kater, M. M., Franken, J., Carney, K. J., Colombo, L., Angenent, G. C. 2001. "Sex Determination in the Monoecious Species Cucumber Is Confined to Specific Floral Whorls." *Plant Cell* 13 (3): 481–93. https://doi.org/10.1105/tpc.13.3.481.

Kenigsbuch, D., Cohen, Y. 1990. "The Inheritance of Gynoecy in Muskmelon." *Genome* 33: 317–20.

Kesh, H., Kaushik, P. 2021. "Advances in Melon (*Cucumis melo* L.) Breeding: An Update." *Scientia Horticulturae* 282: 110045. https://doi.org/10.1016/j.scienta.2021.110045.

Latrasse, David, Rodriguez-Granados, N. Y., Veluchamy, A., Mariappan, K. G., Bevilacqua, C., Crapart, N., Camps, C., Sommard, V., Raynaud, C., Dogimont, C., Boualem, A., Benhamed, M., Bendahmane, A. 2017. "The Quest for Epigenetic Regulation Underlying Unisexual Flower Development in *Cucumis melo*." *Epigenetics & Chromatin* 10: 22. https://doi.org/10.1186/s13072-017-0132-6.

Li, Z., Han, Y., Niu, H., Wang, Y., Jiang, B., Weng, Y. 2020. "Gynoecy Instability in Cucumber (*Cucumis sativus* L.) Is Due to Unequal Crossover at the Copy Number Variation-Dependent Femaleness (*F*) Locus." *Horticulture Research* 7: 32. https://doi.org/10.1038/s41438-020-0251-2.

Loy, J. B. 2006. "Fruit Size in Melon in Monoecious and Andromonoecious Isolines." *Cucurbit Genetics Cooperative Reports* 28–9: 12–3.

Loy, J. B., Natti, T. A., Zack, C. D., Fritts, S. K. 1979. "Chemical Regulation of Sex Expression in a Gynomonoecious Line of Muskmelon." *Journal of the American Society for Horticultural Science* 104 (1): 100–1. https://doi.org/10.21273/JASHS.104.1.100.

Martin, A., Troadec, C., Boualem, A., Rajab, M., Fernandez, R., Morin, H., Pitrat, M., Dogimont, C., Bendahmane, A. 2009. "A Transposon-Induced Epigenetic Change Leads to Sex Determination in Melon." *Nature* 461 (7267): 1135–8. https://doi.org/10.1038/nature08498.

Masuda, K., Akagi, T., Esumi, T., Tao, R. 2019. "Epigenetic Flexibility Underlies Somaclonal Sex Conversions in Hexaploid Persimmon." *Plant and Cell Physiology* 61: 393–402. https://doi.org/10.1093/pcp/pcz207.

Matsumura, H., Miyagi, N., Taniai, N., Fukushima, M., Tarora, K., Shudo, A., Urasaki, N. 2014. "Mapping of the Gynoecy in Bitter Gourd (*Momordica charantia*) Using RAD-Seq Analysis." *PLoS One* 9 (1): e87138. https://doi.org/10.1371/journal.pone.0087138.

Monforte, A. J., Diaz, A., Caño-Delgado, A., van der Knaap, E. 2014. "The Genetics of Fruit Morphology in Horticultural Crops: Lessons from Tomato and Melon." *Journal of Experimental Botany* 65 (16): 4625–37. https://doi.org/10.1093/jxb/eru017.

Murase, K., Shigenobu, S., Fujii, S., Ueda, K., Murata, T., Sakamoto, A., Wada, Y., Yamaguchi, K., Osakabe, Y., Osakabe, K., Kanno, A., Ozaki, Y., Takayama, S. 2017. "MYB Transcription Factor Gene Involved in Sex Determination in Asparagus officinalis." *Genes to Cells: Devoted to Molecular & Cellular Mechanisms* 22 (1): 115–23. https://doi.org/10.1111/gtc.12453.

Nashiki, A., Matsuo, H., Takano, K., Fitriyah, F., Isobe, S., Shirasawa, K., Yoshioka, Y. 2023. "Identification of Novel Sex Determination Loci in Japanese Weedy Melon." *Theoretical & Applied Genetics* 136 (6): 136. https://doi.org/10.1007/s00122-023-04381-3.

Noguera, F. J., Capel, J., Alvarez, J. I., Lozano, R. 2005. "Development and Mapping of a Codominant SCAR Marker Linked to the Andromonoecious Gene of Melon." *Theoretical & Applied Genetics* 110 (4): 714–20. https://doi.org/10.1007/s00122-004-1897-0.

Périn, C., Hagen, L. S., Giovinazzo, N., Besombes, D., Dogimont, C., Pitrat, M. 2002. "Genetic Control of Fruit Shape Acts Prior to Anthesis in Melon (*Cucumis melo* L.)." *Molecular Genetics and Genomics* 266: 933–41. https://doi.org/10.1007/s00438-001-0612-y.

Peterson, C. E., Owens, K. W., Rowe, P. R. 2022. "Wisconsin 998 Muskmelon Germplasm." *HortScience* 18 (1): 116. https://doi.org/10.21273/HORTSCI.18.1.116.

Rashid, D., Devani, R. S., Rodriguez-Granados, N. Y., Abou-Choucha, F., Troadec, C., Morin, H., Tan, F., Marcel, F., Huang, H., Hanique, M., Zhang, S., Verdenaud, M., Pichot, C., Rittener, V., Huang, Y., Benhamed, M., Dogimont, C., Boualem, A., Bendahmane, A. 2023. "Ethylene produced in carpel primordia controls *CmHB40* expression to inhibit stamen development." *Nature Plants*. https://doi.org/10.1038/s41477-023-01511-z.

Renner, S. S. 2014. "The Relative and Absolute Frequencies of Angiosperm Sexual Systems: Dioecy, Monoecy, Gynodioecy, and an Updated Online Database." *American Journal of Botany* 101 (10): 1588–96. https://doi.org/10.3732/ajb.1400196.

Revanasidda, A., Belavadi, V. V. 2019. "Floral Biology and Pollination in *Cucumis melo* L., a Tropical Andromonoecious Cucurbit." *Journal of Asia-Pacific Entomology* 22 (1): 215–25. https://doi.org/10.1016/j.aspen.2019.01.001.

Rodriguez-Granados, N. Y., Ramirez-Prado, J. S., Brik-Chaouche, R., An, J., Manza-Mianza, D., Sircar, S., Troadec, C., Hanique, M., Soulard, C., Costa, R., Dogimont, C., Latrasse, D., Raynaud, C., Boualem, A., Benhamed, M., Bendahmane, A. 2022. "*CmLHP1* Proteins Play a Key Role in Plant Development and Sex Determination in Melon (*Cucumis melo*)." *Plant Journal: For Cell & Molecular Biology* 109 (5): 1213–28. https://doi.org/10.1111/tpj.15627.

Rosa, J. T. 1928. "The Inheritance of Flower Types in *Cucumis* and *Citrullus*." *Hilgardia* 3 (9): 233–50. https://doi.org/10.3733/hilg.v03n09p233.

Sakata, Y., Fukino, N., Ohara, T., Sugiyama, M., Yoshioka, Y. 2013. "Effect of Monoecious Trait on Fruit Shape and Total Soluble Solid Contents of Melon" *Horticultural Research (Japan)* 12: 15–22.

Tsugama, D., Matsuyama, K., Ide, M., Hayashi, M., Fujino, K., Masuda, K. 2017. "A Putative *MYB35* Ortholog Is a Candidate for the Sex-Determining Genes in *Asparagus officinalis*." *Scientific Reports* 7: 41497. https://doi.org/10.1038/srep41497.

Wall, J. R. 1967. "Correlated Inheritance of Sex Expression and Fruit Shape in *Cucumis*." *Euphytica* 16 (2): 199–208. https://doi.org/10.1007/BF00043455.

Wang, Z., Zhang, S., Yang, Y., Li, Z., Li, H., Yu, R., Luan, F., Zhang, X., Wei, C. 2022. "Novel Bisexual Flower Control Gene Regulates Sex Differentiation in Melon (*Cucumis melo* L.)." *Journal of Agricultural & Food Chemistry* 70 (49): 15401–14. https://doi.org/10.1021/acs.jafc.2c05998.

Zhang, S., Tan, F.-Q., Chung, C.-H., Slavkovic, F., Devani, R. S., Troadec, C., Marcel, F., Morin, H., Camps, C., Roldan, M. V. G., Benhamed, M., Dogimont, C., Boualem, A., Bendahmane, A. 2022. "The Control of Carpel Determinacy Pathway Leads to Sex Determination in Cucurbits." *Science* 378 (6619): 543–9. https://doi.org/10.1126/science.add4250.

5 Breeding for Virus Resistance in Melon (*Cucumis melo*)

James D. McCreight, William M. Wintermantel, Shaonpius Mondal, and Prabin Tamang

5.1 INTRODUCTION

Tapley et al. (1937) summarized the extensive knowledge of melon varieties and breeding in the United States, when the emphasis was on novel types and fruit quality. Whitaker and Jagger (1937) stated that "prior to 1850 most of the few varieties of muskmelons then grown in this country [United States] were introduced from Europe." They further noted that "[t]hirty or more years ago growers noticed differential susceptibility of muskmelon to certain diseases," specifically that the variety Hoodoo was mentioned as being resistant to blight. Breeding for resistance to foliar diseases began in 1904 with mass selection for resistance to rust, incited by *Macrosporium cucumerinum* Ell. and Ev. (Blinn, 1908). Melon powdery mildew (*Podosphaera xanthii*)–resistance breeding began shortly after the initial outbreak in Imperial Valley, California, in 1925 (Jagger and Scott, 1937).

In 1941, virus-resistance breeding in melon was prompted after cucumber mosaic virus (CMV) became a production-limiting factor in New York, with a systematic search for genetic resistance in a collection of germplasm from India, Turkey, Afghanistan, Brazil, Paraguay, and Argentina (Enzie, 1943). Since then, an increasing array of viruses has been observed on cucurbits, including melon, and resistance has been found to many of them (Table 5.1).

The state of host plant resistance to 30 viruses in cucurbits was summarized previously by Provvidenti (1993), with the review discussing the state of virus resistance in cucurbits from 1943 through 1993. As with most plants, viruses infecting and causing disease in cucurbit crops vary greatly. Some can be transmitted mechanically, while others can only be transmitted by vector organisms, including aphids, beetles, fungi, leafhoppers, nematodes, thrips, or whiteflies. A few have also been demonstrated to be seedborne in cucurbit crops, including squash mosaic virus (SqMV), melon necrotic spot virus (MNSV), and the cucumber green mottle mosaic virus (CGMMV)–related tobamoviruses. Host plant resistance had been reported for 9 of the 18 viruses known to infect melon at that time (Table 5.2). Resistance to beet pseudoyellows virus (BPYV) had been identified in cross-incompatible wild relatives (Esteva et al., 1989). Potential sources of resistance to lettuce infectious yellows virus

 DOI: 10.1201/9781003715757-5

TABLE 5.1
Numbers of Viruses Infecting Cucurbits and Melon and Numbers of Melon-Infecting Viruses with at Least One Reported Resistance Source, 1993 (Provvidenti, 1993) and 2017 (Keinath et al., 2017)

			Melon		Melon resistance sources	
Source	**Year**	**Cucurbits**	***n***	**%**	***n***	**%**
Provvidenti	1993	30	18	60	9	50
Keinath et al.	2017	52	35	67	16	45
Changes over 24 years		+73%	+94%	+7%	+78%	−5%

(LIYV) in PI 313970 (McCreight, 1992) were not addressed in that review, although it was known at that time.

Provvidenti's 1993 review covered most of the broad array of viruses commonly infecting melon in many areas of the world. There was no source of resistance at that time for CGMMV, but two sources of resistance were reported in 2013 and 2021. No sources of resistance are known for tobacco ringspot virus or tomato ringspot virus.

By 2017, 52 viruses were known to infect cucurbits (Keinath et al., 2017), a 73% increase in 24 years over the preceding 50-year period starting when Enzie (1943) first reported host plant resistance to CMV (Tables 5.1 and 5.2). The number of viruses specifically known to infect melon increased even more sharply, 94%, nearly doubling. In addition, the proportion of viruses for which at least one source of resistance is known increased by 78%. However, taken together, the percentage of known melon-infecting viruses with at least one known source of resistance dropped slightly from 50% in 1993 to 45% in 2017. (Table 5.1).

These numbers do not convey shifts in importance among the viruses, including emergence of novel viruses or new distributions due to changes in melon production patterns, for example, international production and import of melons from Central America and Mexico to the United States, and practices, such as the use and movement of transplants that may introduce novel viruses and their respective vectors into new production areas. Insect-transmitted viruses are greatly influenced by vector populations, which, in turn, are influenced by weather and larger agricultural production changes, including annual variation in cropping patterns that may affect their respective vectors. For example, whitefly (*Bemisia tabaci*) populations increased 300-fold from the mid-1970s to the mid-1980s and 1,600-fold by the mid-1990s, possibly in part due to overuse of synthetic organic pesticides and subsequent resistance to pesticides (Wisler et al., 1998), but also due to displacement of the traditional sweetpotato whitefly (*B. tabaci* biotype A (aka NW1) by the invasive *B. tabaci* biotype B (aka MEAM1) (Cohen et al., 1992). The prevalence of aphid-transmitted viruses can be influenced by winter temperatures and rainfall. This is illustrated by the incidence and severity of the potyviruses, watermelon mosaic virus (WMV) and zucchini yellow mosaic virus (ZYMV) in the Central Valley of California.

TABLE 5.2
Seventeen Melon Viruses and Their Respective Reported Sources of Genetic Resistance or Tolerance

Group	Virus, full name	Resistance source(s)
Begomo	CuLCrV, cucurbit leaf crumple virus	PI 124111, PI 124112, PI 179901, PI 234607, PI 313970, and PI 414723 (McCreight et al., 2008)
Begomo	ToLCNDV, tomato leaf curl New Dehli virus	ND2014–1 V isolate LM47, LM96, LM78, LM77, LM8, LM66, LM98, LM89, LM93, LM60, LM95, LM62, LM52, LM57, LM55, LM75, LM70, LM73 LM91, LM90, (Rosa et al., 2018) Spanish isolate Kharbuja (tolerant), PI 124112, PI 414723 (López et al., 2015) WM9, WM7, Pinonet Piel de sapo (Sáez et al., 2017) ES13–35 isolate AM 87, IC-274014, PI 282448, PI 179901, HSD 2445–005, HSD 2458 B, PI 414723, MR-1, PI 124112, Tibish Kordofan, WM9, WM7, PI 313970 (90625), PI 164723 (Romay et al., 2019)
Begomo	WmCSV, watermelon chlorotic stunt virus	HSD 2445–005, PI 282448, 90625 (PI 313970), PI 124112, and PI 414723 (Yousif et al., 2007)
Yellowing	BPYV, beet pseudoyellows virus	Cross-incompatible wild relatives (Esteva et al., 1989)
Yellowing	CABYV, cucurbit aphid-borne yellows virus	Faizabadi Phoont, 90625 (PI 313970), PI 124112, PI 124440, PI 255478, PI 282448, and PI 414723 PI 124112 (Dogimont et al., 1997, 1996), TGR-1551 (Kassem et al., 2015)
Yellowing	CCYV, cucurbit chlorotic yellows virus	JP 138332, JP 216154, JP 216155, JP 216751, JP 91204 (MR-1) (Okuda et al., 2001)
Yellowing	CYSDV, cucurbit yellow stunting disorder virus	TGR-1551 (Carmen) and TGR-1937 (López-Sesé and Gómez-Guillamón, 2000; Palomares-Rius et al., 2018), PI 313970 (McCreight and Wintermantel, 2011), Ames 20203, PI 614185, and PI 614213 (McCreight and Wintermantel, 2008)
Yellowing	LIYV, lettuce infectious yellows virus	PI 313970 (McCreight, 2000)
Mosaic	CGMMV, cucumber green mottle mosaic virus	Chang Bougi (Sugiyama, 2013), Freeman's Cucumber (Ruiz et al., 2021)
Mosaic	CMV, cucumber mosaic virus	Enzie's Conomon, Makuwa (Karchi et al., 1975; Risser et al., 1977; Takada et al., 1979)
Mosaic	MeSMV, melon severe mosaic virus	Numerous potential sources

Mosaic	PRSV-W, papaya ringspot virus–W MWMV, Moroccan watermelon mosaic virus	PI 180280 and derived lines: B66–5 and WMR 29(Bohn et al., 1980; Webb, 1967), PI 180283 (Pitrat and Lecoq, 1983), PI 124112 (Webb, 1963)
Mosaic	SqMV, squash mosaic	China 51 (Provvidenti, 1998)
Mosaic	WMV, watermelon mosaic virus	Freeman's cucumber (Karchi et al., 1975), PI 414723 (?), TGR-1551 (Diaz-Pendon et al., 2005)
Mosaic	ZYMV, zucchini yellow mosaic virus	PI 414723 strain-specific (Pitrat and Lecoq, 1984), Hannah's Choice F_1 (Henning et al., 2005)
Other	CVYV, cucumber vein yellowing virus	PI 164323(Pitrat et al., 2012)(Pitrat et al., 2012)
Other	MNSV, melon necrotic spot virus	PMR 5, Planter's Jumbo (Coudriet et al., 1981), Doublon (Mallor Gimenez et al., 2003), Cantaloup Haogen (Mallor et al., 2004) (Gonzales-Garza et al., 1979)

The development of virus-resistant melon cultivars has proven to be complex for several reasons including but not limited to strain variation, for example, CMV, virus–vector–host interactions (Keinath et al., 2017; Kishaba et al., 1992; Lecoq et al., 1980; Pitrat and Lecoq, 1980), the increasing frequency of mixed virus infections, such as infection by CMV and potyviruses in the same plants (Giordano et al., 2023), and virus–host interactions that can lead to physiological reactions such as gene silencing (McCreight et al., 2005).

This review focuses on progress in breeding melons for resistance to known viruses.

The melon taxonomy most frequently used in this review follows that of Pitrat (2016), whereby the great phenotypic/horticultural variation in melon is conformed to horticultural rather than botanical classification (Brickell et al., 2004).

One of the early efforts, perhaps the earliest, to breed melon resistant to insect-transmitted viruses was an evaluation of 84 accessions from India, Turkey, Afghanistan, Brazil, Paraguay, and Argentina by Enzie (1943) for resistance to CMV that identified three Group Conomon accessions as potential resistance sources. Inheritance in one of these accessions, which has since been known as Freeman's Cucumber, was reported to be controlled by "three recessive factors" (Karchi et al., 1975). Resistance in PI 161375 ('Songwhan Charmi', Group Chinensis) also proved to also be oligogenic.

A search for resistance to papaya ringspot virus (PRSV), formerly known as watermelon mosaic virus–1 (WMV-1), initiated in 1961, resulted in the identification of resistance in PI 180280 (Group Kachri) and the development of netted cantaloupe breeding lines B66–5 (Webb, 1967) and WMR 29 (Bohn et al., 1980).

WMR 29 originated from a cross of PI 180280 with an F_2 selection from a cross of PI 124111 × Crenshaw. Interestingly, a dried, mature staminate flower of PI 180280 was sent in an envelope from Beltsville, Maryland, to La Jolla, California, where the initial cross was made. This was followed by five direct backcrosses to PMR 29 to combine resistance with desirable fruit quality. Five generations of mass selection improved fruit quality and were followed by two generations of self- and sib-pollination for uniform reaction to PRSV and powdery mildew race 2 (*Podosphaera fuliginaea*). Finally, nine generations of mass selection for fruit quality were done in open-field plantings.

Subsequent development of 'Hannah's Choice F_1' combined resistances in PI 414723 (Group Kachri) to PRSV, watermelon mosaic virus (WMV, formerly known as WMV-2), and ZYMV with eastern U.S. muskmelon-type horticultural qualities (Group Cantalupensis Sub-group American Eastern; Henning et al., 2005; Pitrat, 2016).

An interesting example is the emergence and epidemic spread of LIYV on melon in the lower Sonoran Desert areas of Arizona and California in the United States and Mexico during the early 1980s. After adversely impacting fall season melons for approximately 10 years, it nearly entirely disappeared over a period of months in 1991 with the introduction of the silver leaf whitefly, *Bemisia tabaci* (Gennadius) biotype B (aka MEAM1), in the fall of 1990 (Cohen et al., 1992) and the sudden near-complete loss of its vector, sweet-potato whitefly, *B. tabaci* biotype A (aka NW1). LIYV has not been detected in the desert production region since the early 1990s. Biotype B was a poor transmitter of LIYV but was devastating to the melon plants through a frequently lethal combination of a shorter life cycle and increased phloem sap extraction rates compared with biotype A (Byrne and Bellows, 1991; Costa et al., 1991), often wilting the plants. The higher feeding rate of biotype B increased honeydew excreted on plant and fruit surfaces (Figure 5.1), which served as a medium for sooty mold fungi, for example, *Capnodium* spp., that reduced photosynthesis, further debilitating infested plants. Melons planted for fall harvest in the Sonoran Desert Region emerge and begin flowering when diurnal temperatures range from about 30°C (night) to 45 °C (day), ideal for rapid reproduction of biotype B (Byrne and Bellows, 1991), so that abaxial leaf surfaces are almost entirely white,

FIGURE 5.1 Top: a) Whitefly infestation typical on melon, Imperial Valley. Lower left: b) Honeydew on melon leaves secreted by whitefly (*B. tabaci* MEAM1 [biotype B]) results in shiny, sticky leaves and leads to the development of sooty mold. Lower right: c) Extremely high population of whitefly nymphs on an abaxial leaf surface.

in the Imperial Valley of California under conditions of heavy *B. tabaci* MEAM1 feeding on plants, reactions of the F_1 TGR-1551 × Top Mark (Group Cantalupensis Sub-group American Western), and its derived F_2 and testcross generations indicated recessive control of CYSDV resistance in 2012 (McCreight, unpublished data) and 2013 (McCreight et al., 2017). These results contradicted the initial report of dominant resistance (López-Sesé and Gómez-Guillamón, 2000) based on symptom expression. Interestingly, greenhouse tests in 2012 and 2013 of F_1 TGR-1551 × Bolo de Oro in Malaga, Spain, indicated recessive control resistance based on phenotype (McCreight et al., 2017). RT-PCR values of F_1 TGR-1551 × Bolo de Oro (BO) were, however, slightly lower (indicating higher virus titer) than the mid-parent values in both years, significantly different from both parents in 2012 but not from the susceptible parent in 2013 (McCreight et al., 2017). The intermediate RT-PCR values thus enable the use of the simple backcross procedure (Scully and Federer, 1993) for introgressing resistance rather than the alternate backcross and self (ABS) procedure, appropriate for a recessive trait.

Two analyses of geographically distinct CYSDV isolates by Rubio et al. (2001, 1999) demonstrated low genetic diversity among isolates of CYSDV based on single-strand conformational polymorphism (SSCP) comparisons of the heat shock protein (70 [HSP70]) homolog gene of CYSDV and nucleotide sequence analysis, demonstrating >99% identity for the coat protein gene of most CYSDV isolates. This resulted in a grouping of most CYSDV isolates, including those from Spain and North America, in the same (Western) subpopulation. A separate Eastern subpopulation was determined for isolates closely related to a group originally identified from Saudi Arabia (Rubio et al., 1999).

The contradictory data as to whether resistance in TGR 1551 is recessive or dominant have yet to be adequately explained. Perhaps expression is affected by factors in the genome of the susceptible parent. Dominant resistance was expressed in the cross TGR-1551 × Piel de Sapo, whereas recessive resistance was expressed in crosses of TGR-1551 with 'Dulce', BO, and 'Top Mark'.

CYSDV resistance in PI 313970 was observed in two open-field plantings in the fall of 2006 during the initial outbreak of CYSDV in the Sonoran Desert of the southwestern United States, one in Imperial Valley, California (University of California: Desert Research & Extension Center), and the other Yuma, Arizona (University of Arizona, Yuma Agricultural Center (DREC)), approximately 93 km apart in the Sonoran Desert (McCreight and Wintermantel, 2008). A larger sample (100 lines) of melon germplasm in a replicated field test in 2007 revealed resistance reactions by TGR-1551 from three sources: M.L. Guillamón (Estación Experimental "La Mayora", C.S.I.C., Málaga, Spain), K.C. Crosby (Texas A&M University, College Station), and USDA-GRIN (PI 482420 = TGR-1551), as well as Ames 20203, PI 614185, and PI 614213 (McCreight and Wintermantel, 2008).

The introgression of CYSDV resistance from PI 313970 to Western U.S. shipper–type melons ('Top Mark', 'Impac') and 'Greenflesh Honeydew' (Group Inodorus Subgroup Honeydew) was initiated using the ABS procedure, appropriate for a recessive trait (Scully and Federer, 1993). The selection was based on phenotypic (symptom severity) expression in fall-planted (mid-August water date) open-field tests at DREC, where sweetpotato whitefly populations remained very high. Vegetative cuttings taken from the most resistant segregants approximately 10 weeks post-planting were transplanted to a greenhouse at Salinas, California, for backcrossing (see the earlier discussion) and subsequent selfing through the following summer for the next round of CYSDV-resistance selection. Progress was slow due to the nature of the ABS procedure and the single annual selection cycle. CYSDV symptom severity was correlated with virus titer (J.D. McCreight and W.M. Wintermantel, unpublished).

Although CYSDV is most prevalent and severe in fall-planted melons in the Sonoran Desert region, CYSDV is also present in spring-planted (April) melons in the region, with its emergence and impact in spring melons dependent on whitefly population development in the spring. Although sweetpotato whitefly emerges later in the crop production cycle and populations are lower in the spring than in the fall season, fields remaining by mid-June are generally 100% infected and exhibit prominent CYSDV symptoms by mid-June. TGR-1551 and PI 313970 and several other potential CYSDV-resistance sources exhibited higher mean symptom severity ratings in June 2015 than in previous years, but they were significantly lower ($P >$ .0001) than 'Top Mark' (Table 5.3). Despite this difference, segregating progenies exhibited more uniform disease reactions; that is, they appeared to be more susceptible to CYSDV.

TABLE 5.3
Mean CYSDV Symptom Severity Ratings of 12 Melon Lines, Averaged Across Insecticide Treatment (insecticide regimen and no insecticide) at Three Dates, 2015, University of California, Desert Research and Extension Center, Holtville, Where Symptom Severity Was Evaluated on a Visual Scale from 1 (≤10% foliage symptomatic) to 10 (100%)

		Date (2015)		
Line	*n*	6/12	7/1	7/14
PI 116482	8	1.5	7.6	9.2
PI 122847	8	1.0	4.7	5.8
PI 123689	8	2.0	7.4	7.6
PI 124107	8	1.8	8.6	9.1
PI 145594	8	1.0	5.5	7.2
PI 161375	8	1.2	5.9	8.8
PI 313970	8	1.1	3.4	5.2
PI 414723	8	3.1	8.9	9.4
PI 532841	9	1.0	6.3	7.3
TGR 1551	8	1.0	5.0	6.0
TGR 1937	8	1.0	4.9	6.5
Top Mark	8	1.0	8.8	9.4
ANOVA $P > F$		ns	.0001	ns

ANOVA, analysis of variance; ns, not significant, P = 0.05.

In 2018, a single plant expected to carry CYSDV resistance exhibited an atypical CYSDV symptom. Subsequent studies revealed the symptoms was caused by infection with a closely related virus, cucurbit chlorotic yellows virus (CCYV; genus *Crinivirus*, family Closteroviridae), which had not previously been found in the western hemisphere (Wintermantel et al., 2019). CCYV proved to be uniform throughout the field. Of 23 frozen samples collected between 2014 and 2018, 19 were positive for CCYV, and many contained mixed infections of CCYV with CYSDV and/or the ipomovirus, squash vein yellowing virus (SqVYV; *Ipomovirus*, Potyviridae). All 18 archived samples collected from 2010 to 2013 tested negative for CCYV, but CYSDV, the virus originally identified in the samples, was successfully amplified from these archived extracts. This demonstrated that CCYV had been present in the region since 2014 (Wintermantel et al., 2019).

Thus, the selection for CYSDV resistance in these materials was perhaps ineffective from 2014 to 2018 due to CCYV co-infections. Both viruses produce identical symptoms in melon and most cucurbits.

The introgression of CYSDV resistance from PI 313970 to Western U.S.–type shipping melons under Imperial Valley conditions would have to consider mixed CYSDV and CCYV infections.

5.4 VIRUS DETECTION AND QUANTIFICATION IN BREEDING LINES

The steady emergence of viruses across the southern regions of the United States, from California to Florida, has led to challenges for the evaluation of melon germplasm for resistance to specific viruses. Mixed infections have become common (Mondal et al., 2021a, 2023) (Kavalappara et al., 2021). Some produce similar symptoms, for example, CYSDV, CCYV, and cucurbit aphid-borne yellows virus (CABYV; *Polerovirus*, Luteoviridae), which all produce similar chlorotic mottle and interveinal yellowing symptoms on leaves. Others produce distinct symptoms, but some virus symptoms may mask those caused by other viruses or produce a hybrid symptom, including aspects of symptoms typically produced by both viruses. It can be difficult to determine which virus is primarily responsible for symptoms and whether a plant may exhibit resistance to a particular virus. The frequency of mixed infections in the fields where resistance trials are regularly conducted led to challenges for breeding programs because visual symptoms alone cannot be used to evaluate plants for virus resistance.

In order to facilitate breeding efforts in areas with high virus pressure but an abundance of mixed infections, it was necessary to develop methods that not only determine which viruses are infecting a plant but that also quantify virus titer as a measure of resistance. This is especially important, because if resistance is released commercially, it will need to perform well not only when the target virus occurs as a single infection but also when the plant is challenged by more than one virus, as is now frequently occurring in fields throughout the southern and western United States.

While serological detection of viruses is one of the cost-effective means of determining infection and virus titer, most standard methods cannot simultaneously measure the titers of multiple viruses in single reactions. Serological assays, such as double-antibody sandwich enzyme-linked immunosorbent assay (DAS-ELISA), have been traditionally used to detect many plant viruses; however, some antisera cross-react with related viruses. Some (although not all) antisera against viruses in the genus *Crinivirus*, such as CYSDV or CCYV, can cross-react with related viruses (Wintermantel, unpublished), although there are antisera that are able to effectively and specifically detect CYSDV. CCYV detection using DAS-ELISA comes with potential challenges. A recent study by Kubota et al. (2011) resulted in a heavy chain (VHH) antibody that failed to detect CCYV in indirect ELISA tests, but a dot blot immunoassay was successful in identifying the virus. Similarly, some antisera for detecting cucumber green mottle mosaic virus (CGMMV; *Tobamovirus*, Virgaviridae) are known to cross-react with other viruses in the same genus. If a breeding program intends to use serological detection to determine virus infection or titer, the antiserum should first be tested against not only the target virus but also against other related viruses and host plant material to confirm the antiserum is not only specific for the target virus but also sufficiently sensitive.

At the other end of the detection spectrum is high-throughput sequencing (also known as next-generation sequencing), which can be used to determine the presence of essentially any virus that may be present in a plant sample. While this is useful to determine what viruses are present in a plant, it is not practical for large-scale screening of germplasm for virus resistance.

The abundance of mixed infections in cucurbit fields has led to a need for virus detection assays with the ability to detect or quantify multiple viruses in a single, cost-effective test applicable for breeding programs. A few of these have been developed using RT-PCR and quantitative RT-PCR (RT-qPCR) for differentiation of specific combinations of cucurbit viruses (Jailani et al., 2021; Mondal et al., 2023).

RT-PCR is a molecular technique that can detect RNA viruses. Quantitative RT-PCR (RT-qPCR) is used to quantify RNA virus in test samples. Multiplex single-step RT-PCR and RT-qPCR systems offers the flexibility of detecting and quantifying multiple viruses in a single reaction and bypasses a separate step of producing complementary DNA from RNA. Several previous studies have demonstrated methods that can differentiate CYSDV and CCYV using RT-PCR (Adkins et al., 2008; Kousik and Adkins, 2020; Orfanidou et al., 2019; Wang et al., 2014), but to detect the other viruses, separate reactions were usually performed. A few studies further developed methods to quantify CCYV and CYSDV from mixed infections (Abrahamian et al., 2013; Gil-Salas et al., 2007), but these studies were limited to quantification of two viruses. A recent method developed by Jailani et al. (2021) was able to differentiate CYSDV, CCYV, CuLCrV, and SqVYV using RT-PCR, and another recent study by Mondal et al. (2023) developed a single-step RT-PCR and RT-qPCR system that was able to simultaneously detect and quantify CYSDV, CCYV, SqVYV, and CABYV in a single reaction. Given that these viruses are prevalent throughout the world, this approach is very useful for breeding programs, as well as for cucurbit virus detection.

5.5 APPLICATION OF RT-PCR AND RT-qPCR

Mondal et al. (2023) validated both RT-PCR and RT-qPCR and applied them to study field-collected samples. In 2019, 2020, and 2021, several melon plant samples were collected from the Sonoran Desert melon production regions. All four viruses

the markers cysdv22 (24,613,028 bp), cysdv24 (26,766,636 bp), and cysdv63, which explained 21% to 58% of the variation in symptom evaluation in the three families. These findings confirmed the QTL positions identified in the RIL populations and underscored the need to saturate the region between markers cysdv22 and cysdv24. Furthermore, progeny testing using offspring from a selected BC_3 population allowed the candidate interval to be narrowed to a 700-kb region. The SNP markers identified by Pérez-de-Castro et al. (2020) will be useful for MAS to facilitate introgression of CYSDV resistance into elite cultivars.

Another study aimed to identify the resistance loci and the linked markers in CYSDV-resistant line PI 313970 in California using an $F_{2:3}$ population developed from a cross between CYSDV-resistant PI 313970 and susceptible 'Top Mark' (Tamang et al., 2021). The population was phenotyped in naturally infected open field tests in Imperial Valley, California, in 2018 and 2019. However, the resistant parent PI 313970 and its subsequent $F_{2:3}$ progenies displayed uniform yellowing symptoms due to coinfection with CCYV. RT-qPCR was, therefore, used to estimate the virus titer in each plant of the population, providing phenotypic data for QTL analysis.

In this study, two CYSDV-resistance QTL were identified on chromosomes 3 and 5 in PI 313970 (Figure 5.3). A major QTL on chromosome 5 was identified in approximately the same region in both years, explaining 16% of the titer variation in 2018 (interval S5–21,666,550 to S5–22,718,324) and 35% 2019 (S5–20,833,260 to S5–21,666,437; Figures 5.3 and 5.4). Although the intervals are approximately 1.3 Mb apart, they are close when considering the flanking markers, likely due to year-to-year variation in virus titers in the mapping population. This suggests a single QTL on chromosome 5. These results confirmed the presence of a major

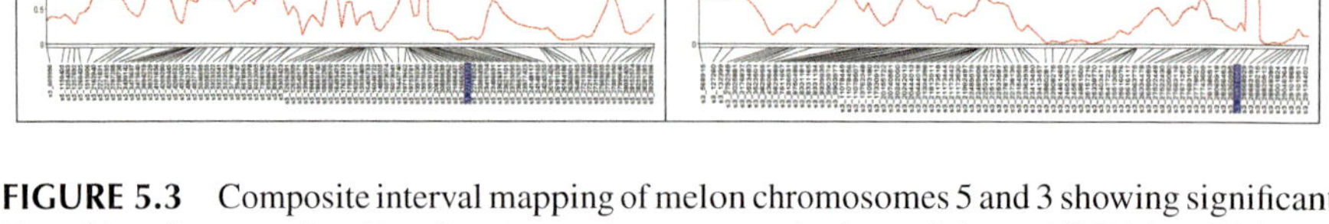

FIGURE 5.3 Composite interval mapping of melon chromosomes 5 and 3 showing significant Cucurbit yellow stunting disorder virus resistance quantitative trait locus (QTL) in a F2:3 Top Mark × PI 313970 population in the fall of 2018. The *x*-axis represents the markers, and the *y*-axis represents the logarithm of the odds (LOD) values. The LOD threshold was calculated by 1000 permutations at $P = .05$ and $P = .01$ significance levels and is denoted by blue and green lines, respectively. The LOD and R^2 values for chromosome 5 QTL are 5.74% and 16%, respectively. The LOD and R^2 values for chromosome 3 QTL are 6.7% and 20%, respectively. The significant marker on chromosome 5 was S5–22217535, which is located at the physical position of 20880639 bp, and the significant marker on chromosome 3 was S3–28571859 located at the physical position of 28571859 (Tamang et al., 2021).

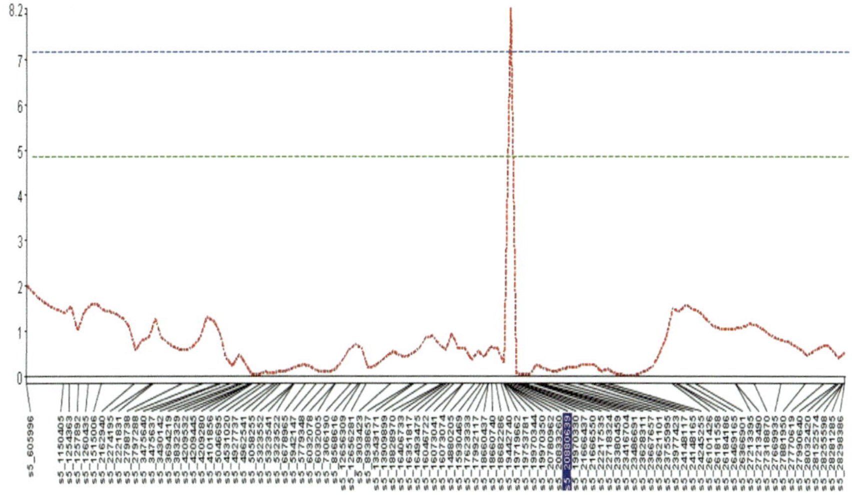

FIGURE 5.4 Composite interval mapping of melon chromosome 5 showing significant cucurbit yellow stunting disorder virus resistance quantitative trait locus (QTL) in a F2:3 'Top Mark' × PI 313970 population, Fall 2019. The *x*-axis represents the markers, and the *y*-axis represents the logarithm of the odds (LOD) values. LOD threshold was calculated by 1000 permutations at $P = .05$ significance level is denoted by a blue line, and $P = .01$ is represented by a green line. The LOD and R^2 values for chromosome 5 QTL were 8.12% and 35%, respectively. The significant marker S5–20880639 is located at 20880639 bp on chromosome 5.

CYSDV-resistance QTL on chromosome 5 using phenotypic severity data (S5–6,810,744 to S5–24,296,585 bp) and virus titer data from RT-qPCR (S5–24,791,006 to S5–27,121,14 bp reported by Pérez-de-Castro et al., 2020).

Markers flanking this chromosome 5 QTL can be valuable for marker-assisted breeding of CYSDV-resistant melons. The QTL region on chromosome 5 (S5–20,833,260 to S5–21,666,437) region contains 96 predicted genes, of which 73 have been annotated. Among these, 12 candidate genes are associated with disease resistance, including genes encoding for serine/threonine-protein kinase-like protein, aldose 1-epimerase family protein, defensin-like protein, clathrin assembly protein, and subtilisin-like protease. Two polymorphic markers, S5–208806396 and S5–21356819, were identified that distinguish the susceptible parent, 'Top Mark', and the resistant parent, PI 313970 (Figure 5.5). These molecular markers were also used to genotype eight resistant F_2 lines whose F2:3 progeny showed low virus titers in 2019. All tested plants matched the resistant parent PI 313970 with these flanking polymorphic markers (Figure 5.6). Further saturation of the chromosome 5 QTL with additional polymorphic markers is essential for fine mapping the CYSDV-resistance gene. These markers will be particularly valuable in breeding for CYSDV resistance even when CCYV is present, as CCYV symptoms are virtually identical to those of CYSDV.

The QTL on chromosome 3, identified at the physical location of 28,571,859 bp (interval S3–28,360,995 to S3–28,568,929), explained 20% of the phenotypic

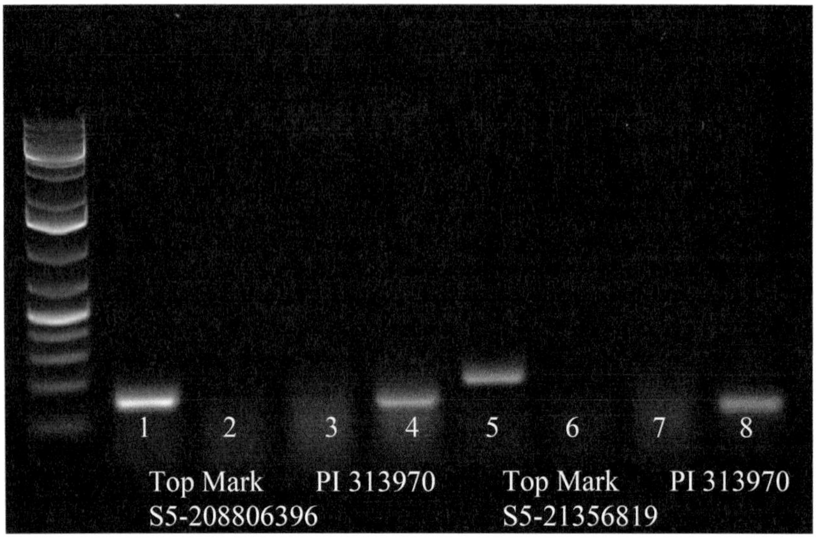

FIGURE 5.5 Genotyping of 'Top Mark' and PI 313970 with two flanking markers (A) S5–20880639 and (B) S5–21356819. Lanes 1, 3, 5, and 7 were 'Top Mark'–specific primers, and lanes 2, 4, 6, and 8 were PI 313970–specific primers. The 'Top Mark'–specific primers amplified only from 'Top Mark' and PI 313970-specific primers amplified only from PI 313970 in both primer sets (Tamang et al., 2021).

variation. This QTL has not been reported previously and may represent a novel CYSDV-resistance gene (Figure 5.3). While it harbors 23 predicted proteins (22 of which are annotated), none are directly linked to disease resistance. As it was detected only with the spring 2018 data, further studies are needed to confirm its significance.

Neither TGR 1551 nor PI 313970 exhibits complete immunity to CYSDV, as RT-qPCR detected low levels of CYSDV in both. Chemical insecticides are widely used to control whiteflies worldwide, but they pose risks to human and environmental health. Host resistance to whitefly could help reduce whitefly populations and consequently decrease chemical usage. Resistance to whitefly may also limit virus spread in the field (Kennedy, 1976). PI 313970 is a potential source of resistance to the whitefly *Bemisia tabaci* (Boissot et al., 2003), although the genetic basis of this resistance remains unknown. Introducing both CYSDV and whitefly resistance from PI 313970 into elite commercial cultivars would be ideal for developing CYSDV-resistant varieties.

For successful cultivation in the southwest U.S. deserts, CYSDV-resistant cultivars must withstand combined pressures from CYSDV, CCYV, and whitefly feeding. Another whitefly-transmitted virus, watermelon chlorotic stunt virus (WmCSV), was identified in 2023 in the Sonoran Desert production region and is now well established and equally prevalent (Wintermantel et al., 2024). While WmCSV is much more severe on watermelon than melon, it does cause foliar symptoms on melon, but the impact on fruit remains to be determined. The genetic complexity of resistance

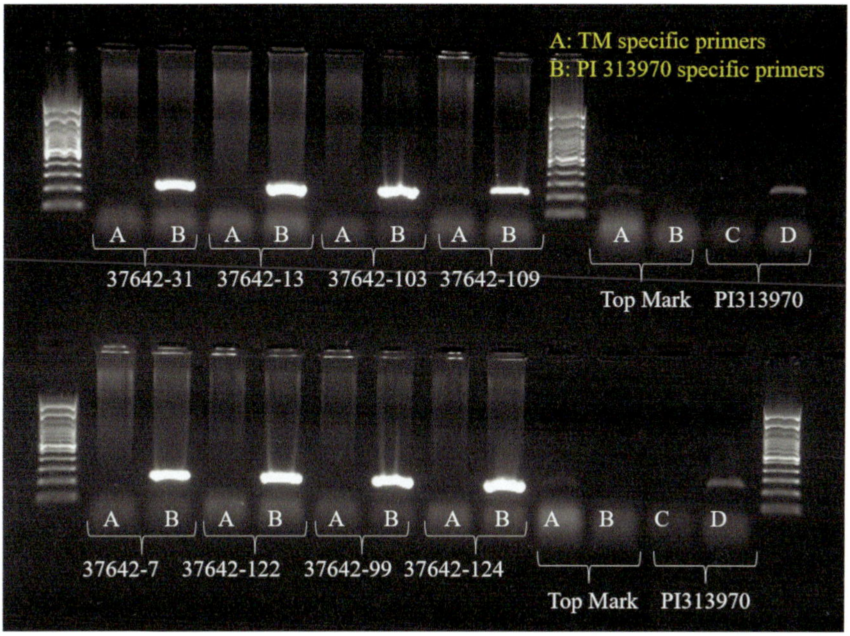

FIGURE 5.6 Genotyping of eight F_2 lines whose $F_{2:3}$ progenies exhibited low virus titer in 2019. All $F_{2:3}$ lines resembled the cucurbit yellow stunting disorder virus–resistant parent PI 313970 for the two flanking markers. (A) 'Top Mark'–specific and (B) PI 313970-specific markers, respectively (Tamang et al., 2021).

and environmental factors can affect the efficacy of resistant cultivars. Continuous research is needed to understand the molecular basis of resistance and develop broad-spectrum resistance against multiple viruses, including CYSDV, CCYV, and WmCSV. Advances in genomic tools, such as genome-wide association studies and CRISPR gene editing, provide new opportunities for enhancing resistance breeding programs. These strategies together can improve the efficiency of developing CYSDV-resistant melon cultivars, which is essential for sustainable agriculture in virus-prone regions.

5.7 CCYV

CCYV is an emerging melon virus and was first reported in Japan in 2004 (Gyoutoku et al., 2009). CCYV was proposed based on the molecular characterization, demonstrating that it is a new species in the genus *Crinivirus* (family Closteroviridae; Gyoutoku et al., 2009). It can infect cucurbits such as melon, cucumber, watermelon, pumpkin, and non-cucurbits crops such as beet, quinoa, datura, and *Nicotiana benthamiana* (Gyoutoku et al., 2009; Okuda et al., 2010). This virus is transmitted by sweetpotato whitefly, *B. tabaci*, and can cause significant yield loss in several melon-growing regions of the world (Gu et al., 2011; Wintermantel et al., 2019).

CCYV is prevalent in East Asia (Chang et al., 2021; Cho et al., 2021; Gu et al., 2011; Huang et al., 2010; Okuda et al., 2010; Zeng et al., 2011), the Middle East (Abrahamian et al., 2013, 2012; Bananej et al., 2013), Europe (Orfanidou et al., 2014), and North Africa (Hamed et al., 2011). CCYV was first reported in the United States in 2019 (Wintermantel et al., 2019). However, it was introduced to the Imperial Valley of California in 2014 but remained undetected until 2019 due to the similar symptoms with CYSDV (Wintermantel et al., 2019).

CCYV induces symptoms on leaves identical to CYSDV, yellowing and interveinal leaf chlorosis, and it is often difficult to visually differentiate them. Infected plants are dwarfed, and there is a significant decrease in the yield and lower sugar content in fruits (Gyoutoku et al., 2009). It is currently controlled by a combination of cultural practices: removing infected plants, using insect-proof net, and chemical treatment against its vector.

There were no melon cultivars resistant to CCYV reported, and no resistance QTL study was performed. Okuda et al. (2013) evaluated 51 melon accessions for CCYV resistance and found only five with no or only faint symptoms: JP 138332, JP 216154, JP 216155, JP 216751 and JP 91204. JP 138332 showed the lowest level of CCYV RNA accumulation compared with those of the other four accessions, which were close to the levels in a susceptible cultivar.

Five potential sources of genetic resistance to CCYV were identified in a survey of 51 melon lines: JP 138332, JP 216154, JP 216155, JP 216751, and JP 91204 (Okuda et al., 2013). JP 216154, of unknown origin, was the most resistant in a follow-up assay based on symptom severity. JP 91204 is the same as MR-1 (Thomas, 1985). Subsequently, JP 138332 had the lowest CCYV content 3 (inoculated 2nd and 3rd leaves and upper 6th or 7th leaf) and 6 (upper 6th or 7th leaf) weeks post-inoculation relative to 'Earl's Seine'. MR-1 had a lower CCYV titer, and PI 313970 had a higher titer than 'Top Mark', the susceptible control in a naturally infected field test (Tamang et al., 2020); however, controlled inoculation of MR-1 in a greenhouse at Salinas found it fully susceptible to a California isolate of CCYV (Wintermantel & McCreight, unpublished).

Kawazu et al. (2018) developed two population maps from the cross between CCYV-resistant parent JP 138332, a snap melon (Group Momordica; Pitrat, 2016), with susceptible parents 'Harkei No. 3' and 'AnMP-5'. The susceptible parents are Earl's Favorite–type melons (Group Inodorus Sub-group Earl's; Pitrat, 2016). 'AnMP-5' possesses resistance to powdery mildew (*P. xanthii*) isolate pxA (Fukino et al., 2008), Fusarium wilt (*Fusarium oxysporum* f. sp. *melonis*; resistance gene *Fom-1*), and cotton-melon aphid (*A. gossypii*). The two F_2 populations were inoculated in a greenhouse using *B. tabaci*. A major QTL on chromosome 1 was detected using these two populations that explained 46.5% to 50% of the phenotypic variation, with a maximum LOD of 12.9. The SSR marker ECM230 was the significant marker associated with this QTL and can be used in the MAS for the breeding CCYV-resistance cultivar. This was the first report of CCYV resistance to date, and the QTL was reported as recessive in accession JP 138332. This QTL may be associated with reduced virus multiplication in the plant.

Kawazu et al. (2024) reported the development of four F_1 hybrid melon lines resistant to CCYV using MAS, CCYV resistance tests, and the selection for fruit quality.

Fruits of these hybrids are globular with yellow-green flesh and whitish pale-green rind covered with dense and thick netting with 15–16°Brix. Resistance to CCYV was reported to be incomplete: The percentage of infected plants was lower, and the symptoms were milder compared with susceptible cultivars. There remains much work to advance CCYV resistance, but this virus continues to emerge as a threat, and its prevalence in association with CYSDV suggests that there is value in advancing resistance to both viruses in the same lines.

5.8 BEGOMOVIRUSES

Begomoviruses: squash leaf curl virus (SLCuV), cucurbit leaf crumple virus (CuLCrV), melon chlorotic leaf curl virus (MCLCuV), watermelon chlorotic stunt virus (WmCSV), tomato leaf curl New Delhi virus (ToLCNDV).

5.8.1 SLCuV, MCLCV, CuLCrV

Squash leaf curl disease was first observed in 1977 causing severe curling on squash plants and mild symptoms on other cucurbit plants, including melon, in the Imperial Valley of California (Flock and Mayhew, 1981). The disease was common in the southwestern U.S. and northern Mexico during the 1980s (Brown et al., 2002). Two strains of SLCuV (formerly SLCV) were eventually associated with this disease, a broad host range strain that can infect melon and other cucurbits, and a strain with a host range restricted to *Cucurbita* spp. (Cohen et al., 1983; Lazarowitz, 1991). Both viruses were originally transmitted by the sweetpotato whitefly (*B. tabaci* NW1; formerly biotype A). With the displacement of *B. tabaci* NW1, the vector of SLCuV, by MEAM1 in 1990, SLCuV gradually disappeared from the southwestern U.S. Recently, however, SLCuV was identified in commercial melon plants collected from Imperial Valley (R.L. Gilbertson, pers. comm.). In single infections SLCuV has limited impact on melon, but studies in Israel demonstrated that coinfection by SLCuV and WmCSV led to significantly reduced plant growth and fruit yield compared to infection with either virus alone (Sufrin-Ringwald and Lapidot, 2011). The potential for synergism between SLCuV and other viruses raises concern that SLCuV may increase in economic importance in melon.

Melon chlorotic leaf curl virus (MCLCuV), which causes a chlorotic leaf curl symptom on melon and reduced fruit production, was first identified in Guatemala (Brown et al., 2007, 2000, 2001; Idris et al., 2008; Maliano et al., 2021), but has not been found outside of Central America to date.

Cucurbit leaf crumple virus (CuLCrV) was initially observed infecting watermelon in 1998 and melon in 1999 in the Imperial Valley, California; Arizona; and Mexico (Brown et al., 2000; Guzman et al., 2000). CuLCrV symptoms on melon are variable and include stunted, curled, and chlorotic tips and leaves with a crumpled appearance (Brown et al., 2000; Guzman et al., 2000; McCreight et al., 2006). The virus was initially described as both cucurbit leaf curl virus (Brown et al., 2000) and cucurbit leaf crumple virus (Guzman et al., 2000) prior to the latter being adopted as the official taxonomic name. CuLCrV is distinct from SLCuV (Flock and Mayhew, 1981) and MCLCuV (Brown et al., 2002, 2001).

Partial resistance to CuLCrV was initially found in melon breeding line MR-1 and six plant introductions – PI 124111 (landrace, India), PI 124112 (landrace, India), PI 179901 (landrace, India), PI 313970 (wild material, India), PI 234607 (landrace, Republic of South Africa), and PI 414723 (cultivated material, India) – in naturally infected field tests and controlled inoculation greenhouse tests (McCreight et al., 2008). PI 236355 was completely resistant in the greenhouse tests and therefore may be a better source for breeding purposes. In addition, it is heterogeneous for exterior fruit characteristics (https://npgsweb.ars-grin.gov/gringlobal/search) that suggest it may represent a variant of European melon germplasm, perhaps Group Cantalupensis (Pitrat, 2016). Resistance in PI 313970 was conditioned by a single recessive gene that appeared allelic to that expressed by the five of the other putative resistance sources; PI 313970 x PI 124112 was not tested (McCreight et al., 2008).

5.8.2 ToLCNDV

ToLCNDV has become a serious concern for the production of melon and other cucurbit crops in Spain and other areas of the Mediterranean basin, where it causes curled leaves and yellow mosaic symptoms on leaves as well as cracked, unmarketable fruit (Siskos et al., 2022) (Font San Ambrosio and Alfaro Fernández, 2015; Juárez et al., 2014). The virus has not been identified in melon production regions of the Western Hemisphere but is of great concern as it would add yet another whitefly-transmitted virus to the ever-increasing virus complex affecting North and Central American cucurbit production. ToLCNDV causes leaf curling, yellow mosaic, and stunting on melon and can cause longitudinal cracking of fruit (Font San Ambrosio and Alfaro Fernández, 2015; Juárez et al., 2014).

Four reports from 2015 to 2019 identified 36 melon accessions resistant/tolerant to four isolates (Table 5.2). Siskos et al. (2022) identified the *C. melo* taxonomies for 26 of the accessions that were not shown in the respective reports (López et al., 2015; Romay et al., 2019; Rosa et al., 2018; Sáez et al., 2017). Virus accumulation was either not detectable or diagnosed at low titers (López et al., 2015; Romay et al., 2019). Kharbuja was tolerant; it exhibited no or mild viral symptoms but had a high virus titer (López et al., 2015; Sáez et al., 2017). PI 164723 and PI 313970 showed a recovery phenotype after infection, and it was confirmed by PCR analysis that the virus titer gradually decreased (Romay et al., 2019). Siskos et al. (2022) summarized susceptible accessions to avoid redundant screening efforts: 76 were susceptible to the isolate ND2014–1 V, 17 to ES13–35.

QTL mapping by Romay et al. (2019) and Sáez et al. (2017) of segregating populations (F2 and BC1) from the cross of ToLCNDV-resistant WM 7 with susceptible 'Piñonet Piel de sapo' (PS), resulted in the identification of one major dominant locus on chromosome 11, consisting of three overlapping QTL (ToLCNDVSy15_11, ToLCNDVSy30_ 11, ToLCNDVVT30_11) and two minor QTL on chromosomes 2 (ToLCNDVVT30_2) and 12 (ToLCNDVSy15_12, ToLCNDVSy30_12, ToLCNDVVT30_ 12; Sáez et al., 2017). The SNP marker D16, which is closely linked to the resistance gene on chromosome 11, can be used in marker-assisted breeding for ToLCNDV resistance in melon (Sáez et al., 2017; Figure 5.2). Two candidate genes, *CmARP4* and *CmNAC*, identified in the WM7–PS material were expressed higher in the inoculated susceptible genotype (PS) compared with the inoculated resistant WM 7 (Román et al., 2019). Accession IC 274014 possessed the highest resistance levels, as all tested individuals were asymptomatic with no detectable virus titer (Román et al., 2019). ToLCNDV resistance is explained by one recessive locus, begomovirus resistance-1 (*bgm*-1), and two dominant genes, begomovirus resistance-2 (*Bgm*-2), and ToLCNDV, according to phenotypic results of F2 population (Román et al., 2019).

See Sisko et al. (2022) for a succinct discussion of patents related to ToLCNDV-resistance and introgression to commercial germplasm, as well as resistances and genes and QTL identified in *Cucurbita moschata* Duchesne and cucumber (*Cucumis sativa* L.) that could facilitate development of resistance in melon.

5.8.3 WmCSV

WmCSV was identified in Mexico in 2017 and in the Sonoran Desert region of California and Arizona in 2023. Although the virus is less severe in melon than watermelon, it causes an irregularly shaped chlorotic or yellow spotting and mosaic symptom that, like other begomoviruses, appears first near the vine ends on infected melon plants (Wintermantel et al., 2024). WmCSV established rapidly in the southwestern U.S. melon production region of Arizona and the Imperial Valley, California, and is now widely prevalent, regularly occurring in mixed infections with CYSDV and CCYV. WmCSV adds to the virus complex in the southwestern United States, and its prevalence is a concern for watermelon production throughout all U.S. cucurbit production regions.

Yousif et al. (2007) evaluated 128 melon accessions, mostly from Sudan, in an open-field test in Sudan under severe whitefly inoculation pressure in order to identify sources of resistance to WmCSV. WmCSV-resistant HSD 2445–005 and 22 previously unscreened accessions were agro-inoculated with an infectious Sudanese clone of WmCSV. Ten accessions exhibited resistance to WmCSV infection upon agro-inoculation. Five of them were confirmed to have complete resistance to WmCSV through graft-inoculation and subsequent field trials in Sudan: Sudanese accession HSD 2445–005, the South African accession PI 282448, and three Indian accessions: 90625 (PI 414723), PI 124112, and PI 414723. Resistant plants exhibited no disease symptoms, and virus DNA could not be detected by hybridization with a WmCSV-specific probe.

5.9 IPOMOVIRUSES: CVYV AND SQVYV

5.9.1 CVYV

CVYV was first described in Israel in the late 1950s (Cohen and Nitzany, 1960) and spread throughout the Mediterranean basin (Lecoq and Desbiez, 2012). It induces mosaic and yellowing symptoms on the leaves and can severely reduce yield and fruit quality. The virus is transmitted in a semi-persistent manner by *Bemisia tabaci* and can also be passed from plant to plant by mechanical inoculation, which facilitates evaluation of plant material for resistance without utilizing a whitefly vector.

Pitrat et al. (2012) evaluated 1188 melon accessions and found PI 164323 (from India) completely resistant to CVYV; it was asymptomatic, and the virus was not detectable in non-inoculated leaves by either DAS-ELISA or RT-PCR. Accession HSD 2458 from Sudan exhibited tolerance; it exhibited very mild mottle, and the virus was detectable in the plant apex by DAS-ELISA or RT-PCR. Furthermore, accession HSD 93–20-A from Sudan exhibited necrotic symptoms followed by rapid death.

Three loci seem to be involved in conditioning resistance to CVYV. The first locus has been tentatively named cucumber vein yellowing resistance (symbol *Cvy-1*); three alleles have been identified: *Cvy-1+* for susceptibility (present in 'Védrantais'), *Cvy-1I* controlling resistance in PI 164323, and *Cvy-12* controlling necrosis in HSD 93–20-A. At an independent second locus, the recessive allele *cvy-2* present in HSD 2458 controls tolerance to the virus. At a third locus, the allele *Cvy-3*, present in 'Ouzbèque 2', controls a highly susceptible type of symptoms in response to CVYV infection.

The development of melon cultivars with durable resistance to CVYV may prove challenging, as Desbiez et al. (2022) demonstrated strain-specific resistance in a study of eight geographically and molecularly diverse isolates (Desbiez et al., 2019). PI 164323 exhibited complete resistance against only two of the eight isolates. Tolerance, as exhibited by accession HSD 2458, was more broadly expressed against six of the eight. Resistance breaking was attributed to a point mutation in the virus genome.

5.9.2 SqVYV

SqVYV, first described from cucurbits in Florida, is named for the yellow vein symptom it produces on leaves of infected squash plants. SqVYV is best known for the severe and relatively unusual symptoms it can produce in watermelon (*Citrullus lanatus*), known as watermelon vine decline, which results in plant death and fruit rind necrosis and discoloration as fields near the harvest stage (Adkins et al., 2008). SqVYV was first observed on *Cucurbita* spp. and melon in the Imperial Valley, California, in 2014 co-infecting with plants along with CYSDV and SLCuV (Batuman et al., 2015). Although the strain present in the Sonoran Desert production region of California and Arizona is genetically divergent from the Florida strain and is capable of inducing vine decline in watermelon under greenhouse conditions, this disease has not been observed in field infections in the western United States (Macedo et al., 2024). SqVYV is closely related to CVYV (Adkins et al., 2007) but does not appear to be of concern for melon production. Melon has not been evaluated for genetic resistance to SqVYV.

5.10 ORTHOTOSPOVIRUSES: MELON SEVERE MOSAIC VIRUS AND MELON YELLOW SPOT VIRUS

5.10.1 Melon Severe Mosaic Virus

Melon severe mosaic virus (MSMV) was first characterized in the state of Guerrero, Mexico, in 2007 causing foliar mosaic and deformation symptoms, as well as fruit splitting of melon plants. MSMV was subsequently found infecting melon, squash/pumpkin (*C. pepo*), cucumber, and watermelon in several states throughout Mexico (Ciuffo et al., 2009), with melon yield losses ranging from 20% to 30%. MSMV is a member of the genus Orthotospovirus and is believed to be transmitted by thrips like other members of the genus; however, no vector has been confirmed to date. MSMV was first observed outside of Mexico on squash (*Cucurbita* spp.) in Arkansas in 2023 associated with foliar mosaic symptoms and fruit distortion (Atanda et al., 2024) and has also been found infecting squash and pumpkin in Mississippi since 2022 associated with mosaic symptoms and frilled edges of leaves (Melanson et al., 2026). Sources of resistance to MSMV have not been reported in melon or other cucurbit crops.

5.10.2 Melon Yellow Spot Virus

Melon yellow spot virus (MYSV) was first observed in 1992 in Japan (Kato et al., 1999); it was subsequently reported in Taiwan (Chao et al., 2010; Peng et al., 2011) and Thailand (Chiemsombat et al., 2008). It is a devastating thrips-transmitted virus of cucurbits. Symptoms in netted Japanese melon (Group Inodorus Sub-group Earl's; Pitrat, 2016) include vein clearing and chlorotic spots on newly developed leaves that progressed to yellowing and necrotic spots; early infection incited mosaic symptoms on fruits, and the disease seriously affected the quality of fruits, including net formation and taste. Sources of resistance have not been reported in melon, although resistance to MYSV has been QTL-mapped in cucumber (Sugiyama et al., 2015).

REFERENCES

Abou-Jawdah, Y., H. Sobh, A. Fayad, H. Lecoq, B. Delécolle, and J. Trad-Ferré. 2000. Cucurbit yellow stunting disorder virus – a new threat to cucurbits in Lebanon. *J. Plant Pathol.* 82:55–60.

Abrahamian, P.E., R. Seblani, H. Sobh, and Y. Abou-Jawdah. 2013. Detection and quantitation of two cucurbit criniviruses in mixed infection by real-time RT-PCR. *J. Virol. Methods* 193:320–326.

Abrahamian, P.E., H. Sobh, and Y. Abou-Jawdah. 2012. First report of cucurbit chlorotic yellows virus on cucumber in Lebanon. *Plant Dis.* 96:1704.

Adkins, S., S.E. Webb, D. Achor, P.D. Roberts, and C.A. Baker. 2007. Identification and characterization of a novel whitefly-transmitted member of the Family Potyviridae Isolated from cucurbits in Florida. *Phytopathology* 97:145–154. https://doi.org/10.1094/phyto-97-2-0145.

Adkins, S., S.E. Webb, C.A. Baker, and C.S. Kousik. 2008. Squash vein yellowing virus detection using nested polymerase chain reaction demonstrates that the cucurbit weed Momordica charantia is a reservoir host. *Plant Dis.* 92:1119–1123. https://apsjournals.apsnet.org/doi/pdf/10.1094/PDIS-92-7-1119.

Atanda, H.Y., S. Smith, J. Pavel, A. Srivastava, B. Osundahunshi, and I.E. Tzanetakis. 2024. First report of melon severe mosaic Orthotospovirus infecting cucurbits in the United States. *Plant Dis.* 108:2246. https://doi.org/10.1094/pdis-02-24-0378-pdn.

Bananej, K., W. Menzel, N. Kianfar, A. Vahdat, and S. Winter. 2013. First report of *Cucurbit chlorotic yellows virus* infecting cucumber, melon and squash in Iran. *Plant Dis.* 97:1005.

Batuman, O., E.T. Natwick, W.M. Wintermantel, T. Tian, J.D. McCreight, L.L. Hladky, and R.L. Gilbertson. 2015. First report of an ipomovirus infecting cucurbits in the Imperial Valley of California. *Plant Dis.* 99:1042. https://doi.org/1010.1094/PDIS-1012-1014-1248-PDN.

Berdiales, B., J.J. Bernal, E. Saez, B. Woudt, F. Beitia, and E. Rodriguez-Cerezo. 1999. Occurrence of *Cucurbit yellow stunting disorder virus* (CYSDV) and *Beet pseudo-yellows virus* in cucurbit crops in Spain and transmission of CYSDV by two biotypes of *Bemisia tabaci*. *Euro. J. Plant Pathol.* 105:211–215.

Blinn, P.K. 1908. Cantaloupe breeding. *Colorado Agric. Exp. Station Bull.* 10.

Bohn, G.W., A.N. Kishaba, and J.D. McCreight. 1980. WMR 29 muskmelon breeding line. *HortScience* 15:539–540.

Boissot, N., D. Lafortune, C. Pavis, and N. Sauvion. 2003. Field resistance to *Bemisia tabaci* in *Cucumis melo*. *HortScience* 38:77–80.

Boissot, N., A. Schoeny, and F. Vanlerberghe-Masutti. 2016. *Vat*, an amazing gene conferring resistance to aphids and viruses they carry: From molecular structure to field effects. *Front. Plant Sci.* https://doi.org/10.3389/fpls.2016.01420.

Brickell, C.D., B.R. Baum, W.L.A. Hetterscheid, A.C. Leslie, J. McNeill, P. Trehane, F. Vrugtman, and J.H. Wiersema. 2004. International code of nomenclature for cultivated plants. *Acta Hort.* 647:9–12, 28–30.

Brown, J.K., J.C. Guerreo, M. Matheron, M. Olsen, and A.M. Idris. 2007. Widespread outbreak of cucurbit yellow stunting disorder virus (CYSDV) in the Sonoran plateau region of the Western U.S. and Pacific coast of Mexico. *Plant Dis.* 91:773.

Brown, J.K., A.M. Idris, C. Alteri, and D.C. Stenger. 2002. Emergence of a new cucurbit-infecting Begomovirus species capable of forming reassortants with related viruses in the *Squash leaf curl virus* cluster. *Phytopathology* 92:734–742.

Brown, J.K., A.M. Idris, M.W. Olsen, M.E. Miller, T. Isakeit, and J. Anciso. 2000. *Cucurbit leaf curl virus*, a new whitefly-transmitted geminivirus in Arizona, Texas, and Mexico. *Plant Dis.* 84:809.

Brown, J.K., A.M. Idris, D. Rogan, M.H. Hussein, and M. Palmieri. 2001. *Melon chlorotic leaf curl virus*, a new begomovirus associated with *Bemisia tabaci* infestation in Guatemala. *Plant Dis.* 85:1027.

Byrne, D.N. and T.S. Bellows. 1991. Whitefly biology. *Annu. Rev. Entomol.* 36:431–457. https://doi.org/10.1146/annurev.en.36.010191.002243.

Célix, A., A. López-Sesé, N. Almarza, M.L. Gómez-Guillamón, and E. Rodríguez-Cerezo. 1996. Characterization of cucurbit yellow stunting disorder virus, a Bemisia tabaci-transmitted closterovirus. *Phytopathology* 86:1370–1376.

Chang, H.Y., L.C. Chen, C.C. Lin, and W.S. Tsai. 2021. First report of cucurbit chlorotic yellows virus infecting melon, watermelon and wild melon in the Philippines. *J. Plant Pathol.* 103:681–682.

Chao, C.H., T.C. Chen, Y.C. Kang, J.T. Li, L.H. Huang, and S.D. Yeh. 2010. Characterization of Melon yellow spot virus infecting cucumber (*Cucumis sativus* L.) in Taiwan. *Plant Pathol. Bull.* 19:41–52.

Chiemsombat, P., O. Gajanandana, N. Warin, R. Hongprayoon, A. Bhunchoth, and P. Pongsapich. 2008. Biological and molecular characterization of tospoviruses in Thailand. *Arch. Virol.* 153:571–577. https://doi.org/10.1007/s00705-007-0024-3.

Cho, I.S., T.B. Kim, J.Y. Yoon, B.N. Chung, J. Hammond, and H.S. Lim. 2021. First report of Cucurbit chlorotic yellows virus infecting *Cucumis melo* (muskmelon and oriental melon) in Korea. *Plant Dis.* 105:2740.

Ciuffo, M., C. Kurowski, E. Vivoda, and B. Copes. 2009. A new Tospovirus sp. in cucurbit crops in Mexico. *Plant Dis.* 93:467–474. https://doi.org/10.1094/PDIS-93-5-0467.

Cohen, S., J.E. Duffus, R.C. Larson, H.Y. Liu, and R.A. Flock. 1983. Purification, serology, and vector relationships of squash leaf curl virus, a whitefly-transmitted geminivirus. *Phytopathology* 73:1669–1673.

Cohen, S., J.E. Duffus, and H.-Y. Liu. 1992. A new *Bemisia tabaci* (Gennadius) biotype in southwestern United States and its role in silverleaf of squash and transmission of lettuce infectious yellows virus. *Phytopathology* 82:86–90.

Cohen, S. and F.E. Nitzany. 1960. A whitefly transmitted virus of cucurbits in israiel. *Beit Dagan, Rehovot-Israel* 44–46.

Costa, H.S., J.K. Brown, and D.M. Byrne. 1991. Life history traits of the whitefly bemisia-tabaci homoptera aleyrodidae on six virus-infected or healthy plant species. *Environ. Entomol.* 20(4):1102–1107. https://doi.org/10.1093/ee/20.4.1102.

Coudriet, D.L., A.N. Kishaba, and G.W. Bohn. 1981. Inheritance of resistance to *Muskmelon necrotic spot virus* in a melon aphid-resistant breeding line of muskmelon. *J. Am. Soc. Hort. Sci.* 106:789–791.

Desbiez, C., P. Caciagli, C. Wipf-Scheibel, P. Millot, L. Ruiz, D. Marian, G. Dafalla, and H. Lecoq. 2019. Evidence for long-term prevalence of cucumber vein yellowing virus in Sudan and genetic variation of the virus in Sudan and the Mediterranean basin. *Plant Pathol.* 68:1268–1275. https://doi.org/10.1111/ppa.13055.

Desbiez, C., M.L. Domingo-Calap, M. Pitrat, C. Wipf-Scheibel, G. Girardot, I. Ferriol, J.J. Lopez-Moya, and H. Lecoq. 2022. Specificity of Resistance and Tolerance to Cucumber Vein Yellowing Virus in Melon Accessions and Resistance Breaking with a Single Mutation in VPg. *Phytopathology* 112:1185-1191. https://doi.org/10.1094/phyto-06-21-0263-r.

Desbiez, C., H. Lecoq, S. Aboulama, and M. Peterschmitt. 2000. First report of *Cucurbit yellow stunting disorder virus* in Morocco. *Plant Dis.* 84:596.

Diaz-Pendon, J.A., R. Fernandez-Munoz, M.L. Gomez-Guillamon, and E. Moriones. 2005. Inheritance of resistance to Watermelon mosaic virus in *Cucumis melo* that impairs virus accumulation, symptom expression, and aphid transmission. *Phytopathology* 840–846.

Dogimont, C., A. Bendahmane, M. Pitrat, E. Burget-Bigeard, L. Hagen, A. Le Menn, J. Pauquet, P. Rouselle, M. Caboche, and C.V. 2004. *New polynucleotide implicated in plant resistance, useful for producing transgenic plants resistant to Aphis gossypii and associated viral transmission, also encoded protein.* World patent WO2004072109-A.

Dogimont, C., A. Bussemakers, J. Martin, S. Slama, H. Lecoq, and M. Pitrat. 1997. Two complementary recessive genes conferring resistance to *Cucurbit aphid borne yellows luteovirus* in an Indian melon line (*Cucumis melo* L.). *Euphytica* 96:391–395.

Dogimont, C., S. Slama, J. Martin, H. Lecoq, and M. Pitrat. 1996. Sources of resistance to cucurbit aphid-borne yellows luteovirus in a melon germ plasm collection. *Plant Dis.* 80:1379–1382.

Enzie, W.D. 1943. A source of muskmelon mosaic resistance found in the oriental pickling melon, *Cucumis melo* var. *conomon*. *Proc. Am. Soc. Hort. Sci.* 43:195–198.

Esteva, J., F. Nuez, and M.L. Gomez-Guillamon. 1989. Resistance to yellowing disease in muskmelon. *Cucurbit Genet. Coop. Rpt.* 12:44–45.

Flock, R.A. and D.E. Mayhew. 1981. Squash leaf curl, a new disease of cucurbits in California. *Plant Dis.* 65:75–76.

Font San Ambrosio, M.I. and A.O. Alfaro Fernández. 2015. El virus de Nueva Delhi (Tomato leaf curl New Delhi virus, ToLCNDV) amplı´a su gama de hospedantes en los cultivos espan˜oles. *Phytoma* 272:25–30.

Fukino, N., T. Ohara, A.J. Monforte, M. Sugiyama, Y. Sakata, M. Kunihisa, and S. Matsumoto. 2008. Identification of QTLs for resistance to powdery mildew and SSR markers diagnostic for powdery mildew resistance genes in melon (*Cucumis melo* L.). *Theor. Appl. Genet.* 118:165–175.

Gilbertson, R.L., E.T. Natwick, T. Turini, and K.S. Mayberry. 2000. Cucurbit leaf crumple, a new geminivirus. *Plant Dis.* 84: (submitted).

Gil-Salas, F.M., J. Morris, A. Colyer, G. Budge, N. Boonham, I.M. Cuadrado, and D. Janssen. 2007. Development of real-time RT-PCR assays for the detection of Cucumber vein yellowing virus (CVYV) and Cucurbit yellow stunting disorder virus (CYSDV) in the whitefly vector Bemisia tabaci. *J. Virol. Methods* 146:45–51.

Giordano, A., I. Ferriol, J.J. López-Moya, and A.M. Martín-Hernández. 2023. cmv1-Mediated resistance to CMV in melon can be overcome by mixed infections with potyviruses. *Viruses* 15:1792.

Gonzales-Garza, R., D.J. Gumpf, A.N. Kishaba, and G.W. Bohn. 1979. Identification, seed transmission; host range pathogenicity of a California isolate of melon necrotic spot virus. *Phytopathology* 69:340–345.

Gu, Q.S., Y.H. Liu, Y.H. Wang, and W.G. Huangfu. 2011. First report of *Cucurbit chlorotic yellows virus* in cucumber, melon, and watermelon in China. *Plant Dis.* 95:73.

Guzman, P., M.R. Sudarshana, Y.S. Seo, M.R. Rojas, E. Natwick, T. Turini, K. Mayberry, and R.L. Gilbertson. 2000. A new bipartite geminivirus (Begomovirus) causing leaf curl and crumpling in cucurbits in the Imperial Valley of California. *Plant Dis.* 84:488. https://doi.org/10.1094/PDIS.2000.84.4.488C.

Gyoutoku, Y., S. Okazaki, A. Furuta, T. Etoh, M. Mizobe, K. Kuno, S. Hayashida, and M. Okuda. 2009. Chlorotic yellows disease of melon caused by Cucurbit chlorotic yellows virus, a new crinivirus. *Japanese J. Phytopathol.* 75:109–111.

Hagen, C., M.R. Rojas, M.R. Sudarshana, B. Xoconostle-Cazares, E.T. Natwick, T.A. Turini, and R.L. Gilbertson. 2008. Biology and molecular characterization of cucurbit leaf crumple virus, an emergent cucurbit-Infecting Begomovirus in the Imperial Valley of California. *Plant Dis.* 92:781–793. https://doi.org/10.1094/PDIS-92-5-0781. PMID: 30769582.

Hamed, K., W. Menzel, G. Dafalla, A.M.A. Gadelseed, and S. Winter. 2011. First report of *cucurbit chlorotic yellows virus* infecting muskmelon and cucumber in Sudan. *Plant Dis.* 95:1321.

Hassan, A.A., H.H. Al-Masri, U.A. Obaji, M.S. Wafi, and N.E. Quronfilah. 1991. Screening of domestic and wild *Cucumis melo* germplasm for resistance to the yellow-stunting disorder in the United Arab Emirates. *Cucurbit Genet. Coop. Rpt.* 14:56–58.

Hassan, A.A. and J.E. Duffus. 1990. Observations and investigations on the yellow and stunting disorder of cucurbits in the United Arab Emirates–a review. *Emirates J. Agr. Sci.* 2:1–16.

Henning, M.J., H.M. Munger, and M.M. Jahn. 2005. 'Hannah's Choice F1': A new muskmelon hybrid with resistance to powdery mildew, Fusarium race 2, and potyviruses. *HortScience* 40:492–493.

Huang, L.H., H.H. Tseng, J.T. Li, and T.C. Chen. 2010. First Report of cucurbit chlorotic yellows virus Infecting Cucurbits in Taiwan. *Plant Dis.* 94:1168. https://doi.org/10.1094/PDIS-94-9-1168B.

Idris, A.M., K. Mills-Lujan, K. Martin, and J.K. Brown. 2008. Melon chlorotic leaf curl virus: Characterization and differential reassortment with closest relatives reveal adaptive virulence in the squash leaf curl virus clade and host shifting by the host-restricted bean calico mosaic virus. *J. Virol.* 82:1959–1967. https://doi.org/10.1128/JVI.01992-07. Epub 2007 Dec 5. PMID: 18057231; PMCID: PMC2258725.

Jagger, I.C. and G.W. Scott. 1937. Development of powdery mildew resistant cantaloup No. 45. *USDA Circ.* 441:6.

Jailani, A.A.K., K. Hendricks, P.D. Roberts, and M.L. Paret. 2021. Development of a simple one-step multiplex RT-PCR system for simultaneous detection of DNA and RNA viruses of cucurbit leaf crumple virus, cucurbit yellow stunting disorder virus, squash vein yellowing virus, and cucurbit chlorotic yellows virus. *Physiol. Mol. Plant Pathol.* 116:101734. https://doi.org/10.1016/j.pmpp.2021.101734.

Juárez, M., R. Tovar, E. Fiallo-Olivé, M.A. Aranda, B. Gosálvez, P. Castillo, E. Moriones, and J. Navas-Castillo. 2014. First detection of tomato leaf curl New Delhi virus infecting zucchini in Spain. *Plant Dis.* 98:857–858. https://doi.org/10.1094/pdis-10-13-1050-PDN.

Kao, J., L. Jia, T. Tian, L. Rubio, and B.W. Falk. 2000. First report of Cucurbit yellow stunting disorder virus (genus Crinivirus) in North America. *Plant Dis*. 84:101.

Karchi, Z., S. Cohen, and A. Govers. 1975. Inheritance of resistance to cucumber mosaic virus in melons. *Phytopathology* 65:479–481.

Kassem, M.A., B. Gosalvez, E. Garzo, A. Fereres, M.L. Gómez-Guillamón, and M.A. Aranda. 2015. Resistance to Cucurbit aphid-borne yellows virus in Melon Accession TGR-1551. *Phytopathology* 105:1389–1396. https://doi.org/10.1094/phyto-02-15-0041-r.

Kato, C., K. Hanada, and M. Kameya-Iwaki. 1999. Transmission mode, host range and electron microscopy of a pathogen causing a new disease of melon (*Cucumis melo*) in Japan. *Japanese J. Phytopathol*. 65:624–627. https://doi.org/10.3186/jjphytopath.65.624.

Kavalappara, S., S. Raman, H. Milner, N.C. Konakalla, K. Morgan, A.N. Sparks, C. McGregor, A.K. Culbreath, W.M. Wintermantel, and S. Bag. 2021. High Throughput Sequencing-Aided Survey Reveals Widespread Mixed Infections of Whitefly-Transmitted Viruses in Cucurbits in Georgia, USA. *Viruses* 13:988. https://doi.org/10.3390/v13060988.

Kawazu, Y., K. Shimomura, S. Maeda, Y. Yamato, S. Ueda, S. Okuda, M. Okuda, and M. Sugiyama. 2018. QTL mapping for resistance to *Cucurbit chlorotic yellows virus* in melon (*Cucumis melo* L.). *Euphytica* 214:239. https://doi.org/10.1007/s10681-018-2321-1.

Kawazu, Y., M. Sugiyama, K. Shimomura, S. Ueda, S. Watanabe, Y. Nakai, S. Maeda, T. Yumbe, R. Yano, T. Yamamoto, and M. Harada. 2024. Development of melon F 1 hybrids with resistance to cucurbit chlorotic yellows virus. *Acta Hort*. https://doi.org/10.17660/ActaHortic.2024.1404.131.

Keinath, A.P., W.M. Wintermantel, and T.A. Zitter. 2017. *Compendium of Cucurbit Diseases and Insect Pests*, 2nd ed. APS Press, St. Paul, Minn.

Kennedy, G.G. 1976. Host plant resistance and the spread of plant viruses. *Environ. Entomol*. 5:827–832.

Kishaba, A.N., S. Castle, D.L. Coudriet, J.D. McCreight, and G.W. Bohn. 1992. Virus transmission by *Aphis gossypii* Glover to aphid-resistant and susceptible muskmelons. *J. Am. Soc. Hort. Sci*. 117:248–254.

Kousik, C.S. and S. Adkins. 2020. Detection of cucurbit yellow stunting disorder virus infecting watermelon in South Carolina. *Plant Health Progress* 21. https://doi.org/10.1094/PHP-03-20-0016-BR.

Kubota, K., T. Usugi, and S. Tsuda. 2011. Production of antiserum and immunodetection of Cucurbit chlorotic yellows virus, a novel whitefly-transmitted crinivirus. *J. Gen. Plant Pathol*. 77:116–120.

Kuo, Y.-W., M.R. Rojas, R.L. Gilbertson, and W.M. Wintermantel. 2007. First report of *Cucurbit yellow stunting disorder virus* in California and Arizona, in association with *Cucurbit leaf crumple virus* and *Squash leaf curl virus*. *Plant Dis*. 91:330. https://doi.org/10.1094/PDIS-91-3-0330B.

Lazarowitz, S.G. 1991. Molecular characterization of two bi-partite geminiviruses causing squash leaf curl disease: role of viral replication and movement functions in determining host range. *Virology* 180:70–80.

Lecoq, H. and C. Desbiez. 2012. Viruses of cucurbit crops in the Mediterranean region: An ever-changing picture. *Adv. Virus Res*. 84:67–126. https://doi.org/10.1016/B978-0-12-394314-9.00003-8.

Lecoq, H., G. Labonne, and M. Pitrat. 1980. Specificity of resistance to virus transmission by aphids in *Cucumis melo*. *Ann. Phytopathol*. 12:139–144.

López, C., M. Ferriol, and M.B. Picó. 2015. Mechanical transmission of tomato leaf curl New Delhi virus to cucurbit germplasm: Selection of tolerance sources in *Cucumis melo*. *Euphytica* 204:679–691. https://doi. org/10.1007/s10681-015-1371-x.

López-Sesé, A.I. and M.L. Gómez-Guillamón. 2000. Resistance to *Cucurbit yellowing stunting disorder virus* (CYSDV) in *Cucumis melo* L. *HortScience* 35:110–113.

Louro, D., M. Vicente, A.M. Vaira, G.P. Accotto, and G. Nolasco. 2000. *Cucurbit yellow stunting disorder virus* (Genus Crinivirus) associated with the yellowing disease of cucurbit crops in Portugal. *Plant Dis.* 84:1156.

Macedo, M.A., T.A. Melgarejo, M. Vasquez-Mayorga, M. Cespedes, M.R. Rojas, T.A. Turini, O. Batuman, W.M. Wintermantel, and R.L. Gilbertson. 2024. Squash vein yellowing virus from California emerged in the Middle East via intragenic and intergeneric recombination events in the hypervariable potyvirus P1 and ipomovirus P1a genes. *J. Gen. Virol.* 105. https://doi.org/10.1099/jgv.0.002033. PMID: 39418084.

Maliano, M.R., T.A. Melgarejo, M.R. Rojas, N. Barboza, and R.L. Gilbertson. 2021. The Begomovirus species melon chlorotic leaf curl virus is composed of two highly divergent strains that differ in their genetic and biological properties. *Plant Dis.* 105:3162–3170. https://doi.org/10.1094/PDIS-08-20-1759-RE. Epub 2021 Nov 7. PMID: 33591835.

Mallor, C., J.M. Alvarez, and M. Luis-Arteaga. 2004. Behaviour of *Cucumis melo* 'Cantaloup Haogen' against melon necrotic spot virus (MNSV), p. 203–208. In: A. Lebeda and H.S. Paris (eds.). *Progress in cucurbit genetics and breeding research. Proceedings of Cucurbitaceae 2004, the 8th EUCARPIA Meeting on Cucurbit Genetics and Breeding, Olomouc, Czech Republic, 12–17 July, 2004.* Palacky University in Olomouc, Olomouc,

Mallor Gimenez, C., J.M. Alvarez Alvarez, and M. Luis Arteaga. 2003. Inheritance of resistance to systemic symptom expression of melon necrotic spot virus (MNSV) in *Cucumis melo* L. 'Doublon'. *Euphytica* 134:319–324.

Masegosa, J.R., C. Martínez, E. Aguado, A. García, G. Cebrián, J. Iglesias-Moya, H.S. Paris, and M. Jamilena. 2020. Response of cucurbita spp. to tomato leaf curl New Delhi Virus inoculation and identification of a dominant source of resistance in *Cucurbita moschata Plant Pathol.* https://doi.org/10.1111/ppa.13268.

McCreight, J.D. 1992. Screening for *Lettuce infectious yellows virus* resistance in melon, p. 160–162. In: R.W. Doruchowski, E. Kozik, and K. Niemirowicz-Szczytt (eds.). *Fifth Eucarpia Cucurbitaceae Symposium, July 27–31, 1992*, Research Institute of Vegetable Crops Skierniewice and Warsaw University of Agriculture Warsaw, Poland.

McCreight, J.D. 2000. Inheritance of resistance to lettuce infectious yellows virus in melon. *HortScience* 35:1118–1120.

McCreight, J.D., H.-Y. Liu, and T.A. Turini. 2005. Resistance to *Cucurbit leaf crumple virus* in melon. *HortScience* 40:122–126.

McCreight, J.D., H.-Y. Liu, and T.A. Turini. 2006. Reaction of Melon PI 313970 to *Cucurbit leaf crumple virus*, p. 139–145. In: G.J. Holmes (ed.). *Proceedings of Cucurbitaceae 2006.* Universal Press, Raleigh, North Carolina.

McCreight, J.D. and W.M. Wintermantel. 2008. Potential new sources of genetic resistance in melon to *Cucurbit yellow stunting disorder virus*, p. 173–179. In: M. Pitrat (ed.). *Cucurbitaceae 2008, Proceedings of the IXth EUCARPIA meeting on genetics and breeding of Cucurbitaceae, May 21–24th, 2008.* INRA, Avignon (France).

McCreight, J.D. and W.M. Wintermantel. 2011. Genetic resistance in melon PI 313970 to *Cucurbit yellow stunting disorder virus. HortScience* 46:1582–1587.

McCreight, J.D., H.-Y. Liu, and T.A. Turini. 2008. Genetic resistance to cucurbit leaf crumple virus in melon. *HortScience* 43:122–126.

McCreight, J.D., W.M. Wintermantel, and E.T. Natwick. 2015. Inheritance of resistance to *Cucurbit yellow stunting disorder virus* (CYSDV) in melon (*Cucumis melo* L.) accessions PI 482431 (TGR 1937) and PI 614479. *In: Proceedings IXVIII. International Plant Protection Congress, 24–27 August 2015*, Berlin.

McCreight, J.D., W.M. Wintermantel, E.T. Natwick, J.W. Sinclair, K.M. Crosby, and M.L. Gomez-Guillamon. 2017. Recessive resistance to *Cucurbit yellow stunting disorder virus* in melon TGR 1551. *Acta Hort.* 1151:101–107.

Melanson, R.A., Chen, C., Stevens, K.A., Jenkins Hladky, L., Al Rwahnih, M., and Wintermantel, W.M. 2026. Identification of melon severe mosaic virus in cucurbits in Mississippi

highlights the need for routine virus monitoring with assays targeting multiple viruses. *Plant Disease* (Accepted with Revision [final revisions submitted]).
Mondal, S., L.J. Hladky, P.L. Fashing, J.D. McCreight, T.A. Turini, and W.M. Wintermantel. 2021a. First report of cucurbit yellow stunting disorder virus and cucurbit chlorotic yellows virus in melon in the Central Valley of California. *Plant Dis.* https://doi.org/10.1094/PDIS-01-21-0184-PDN.
Mondal, S., L.J. Hladky, R.A. Melanson, R. Singh, E. Sikora, and W.M. Wintermantel. 2021b. First report of cucurbit yellow stunting disorder virus and cucurbit chlorotic yellows virus in cucurbit crops in Alabama. *Plant Dis.* https://doi.org/10.1094/PDIS-05-21-0922-PDN.
Mondal, S., L.J. Hladky, and W.M. Wintermantel. 2023. Differential seasonal prevalence of yellowing viruses infecting melon crops in Southern California and arizona determined by multiplex RT-PCR and RT-qPCR. *Plant Dis.* 107:2653–2664. https://doi.org/10.1094/PDIS-06-22-1512-RE. Epub 2023 Sep 11. PMID: 36723958.
Okuda, M., S. Okazaki, S. Yamasaki, S. Okuda, and M. Sugiyama. 2010. Host range and complete genome sequence of *Cucurbit chlorotic yellows virus*, a new member of the genus Crinivirus. *Phytopathology* 560–566.
Okuda, S., M. Okuda, M. Sugiyama, Y. Sakata, M. Takeshita, and H. Iwai. 2013. Resistance in melon to *Cucurbit chlorotic yellows virus*, a whitefly-transmitted crinivirus. *Euro. J. Plant Pathol.* 135:313–321.
Okuda, M., S. Taba, S. Tsuda, S. Hidaka, M. Kameya-Iwaki, and K. Hanada. 2001. Comparison of the S RNA segments among Japanese isolates and Taiwanese isolates of watermelon silver mottle virus. *Arch. Virol.* http://www.springer.co.at. ISSN: 0304–8608 146:389–394.
Orfanidou, C.G., V.I. Maliogka, and N.I. Katis. 2014. First report of Cucurbit chlorotic yellows virsu in cucumber, melon, and watermelon in Greece. *Plant Dis.* 98:1446.
Orfanidou, C.G., L.C. Papayiannis, P.G. Pappi, N.I. Katis, and V.I. Maliogka. 2019. Criniviruses associated with cucurbit yellows disease in Greece and Cyprus: An ever-changing scene. *Plant Pathol.* 68:764–774. https://doi.org/10.1111/ppa.12986.
Palomares-Rius, F.J., A. Garces-Claver, M. Belen Pico, C. Esteras, F. Juan Yuste-Lisbona, and M. Luisa Gomez-Guillamon. 2018. 'Carmen', a Yellow Canary melon breeding line resistant to Podosphaera xanthii, Aphis gossypii, and Cucurbit yellow stunting disorder virus. *Hortscience* 53:1072–1075. https://doi.org/10.21273/HORTSCI13013–18.
Peng, J.-C., S.D. Yeh, L.-H. Huang, J.T. Li, Y.-F. Cheng, and T.C. Chen. 2011. Emerging threat of thrips-borne *Melon yellow spot virus* on melon and watermelon in Taiwan. *Euro. J. Plant Pathol.* 130:205–214.
Pérez-de-Castro, A., M. López-Martín, C. Esteras, A. Garcés-Claver, F.J. Palomares-Ríus, M.B. Picó, and M.L. Gómez-Guillamón. 2020. Melon genome regions associated with TGR-1551-derived resistance to *Cucurbit yellow stunting disorder virus*. *Int. J. Mol. Sci.* 21:5970. https://doi.org/10.3390/ijms21175970.
Pitrat, M. 2016. Melon genetic resources: Phenotypic diversity and horticultural taxonomy. In: R. Grumet, N. Katzir, and J. Garcia-Mas (eds.). *Genetics and Genomics of Cucurbitaceae*. Springer Science+Media, New York, p. 25–60. https://doi.org/10.1007/7397_2016_10.
Pitrat, M. and H. Lecoq. 1980. Inheritance of resistance to cucumber mosaic virus transmission by *Aphis gossypii* in *Cucumis melo*. *Phytopathology* 70:958–961.
Pitrat, M. and H. Lecoq. 1983. Two alleles for watermelon mosaic virus 1 resistance in muskmelon. *Cucurbit Genet. Coop. Rpt.* 6:52–53.
Pitrat, M. and H. Lecoq. 1984. Inheritance of zucchini yellow mosaic virus resistance in *Cucumis melo* L. *Euphytica* 33:57–61. https://doi.org/10.1007/BF00022750.
Pitrat, M., C. Wipf-Scheibel, D. Besombes, C. Desbiez, and H. Lecoq. 2012. Resistance of melon to cucumber vein yellowing virus (CVYV). In: N. Sari, I. Solmaz, and V. Aras (eds.). *Cucurbitaceae 2012, Proceedings of the Xth EUCARPIA Meeting on Genetics and Breeding of Cucurbitaceae, Antalya (Turkey), October 15–18th, 2012*, p. 157–164. Agricultural and Food Sciences, Biology.

Pitter, P.L., S. Mondal, P.G. Chang, L. Myers Morgan, S. Aikman, W.M. Wintermantel, and P.F. Tennant. 2024. First report of cucurbit yellow stunting disorder virus infecting cucurbit crops in Jamaica. *Plant Dis.* https://doi.org/10.1094/PDIS-08-23-1551-PDN. Epub ahead of print. PMID: 38414194.

Provvidenti, R. 1993. Resistance to viral diseases of cucurbits. In: M.M. Kyle (ed.). *Resistance to Viral Diseases of Vegetables*. Timber Press, Portland, Ore, p. 8–43.

Provvidenti, R. 1998. A source of a high level of tolerance to squash mosaic virus in a melon-from China. *Cucurbit Genet. Coop. Rpt.* 29–30.

Risser, G., M. Pitrat, and J.C. Rode. 1977. Etude de la resistance du melon (*Cucumis melo* L.) au virus de la mosaique du concombre. *Annals de l'Amelioration des Plantes* 27:509–522.

Román, B., P. Gómez, D. Janssen, B. Picó, and C. López. 2019. Candidate gene analysis of Tomato leaf curl New Delhi virus resistance in Cucumis melo. *Scientia Hort.* 243:12–20. https://doi.org/10.1016/j.scienta.2018.07.005.

Romay, G., M. Pitrat, H. Lecoq, C. Wipf-Scheibel, P. Millot, G. Girardot, and C. Desbiez. 2019. Resistance against melon chlorotic mosaic virus and tomato leaf curl New Delhi Virus in melon. *Plant Dis.* 103:2913–2919. https://doi.org/10.1094/PDIS-02-19-0298-RE.

Rosa, E., M.R. Figàs, C. Casanova, M. Hurtado, J.M. González-Cabezuelo, J. Prohens, and S. Soler. 2018. Identificación de tres fuentes de resistencia al virus del rizado del tomate de Nueva Delhi (ToLCNDV) en melón. *ACTAS de Horticultura* 80:72.

Rubio, L., Y. Abou-Jawdah, H.-X. Lin, and B.W. Falk. 2001. Geographically distant isolates of the crinivirus *Cucurbit yellow stunting disorder virus* show very low genetic diversity in the coat protein gene. *J. Gen. Virol.* 82:929–933.

Rubio, L., J. Soong, J. Kao, and B.W. Falk. 1999. Geographic distribution and molecular variation of isolates of three whitefly-borne closteroviruses of cucurbits: *Lettuce infectious yellows virus*, *Cucurbit yellow stunting disorder virus*, and *Beet pseudo-yellows virus*. *Phytopathology* 89:707–711.

Ruiz, L., C. López, B. Picó, and D. Janssen. 2021. Resistance to cucumber green mottle mosaic virus in *Cucumis melo*. *Plants* 10:1077. https://doi.org/10.3390/plants10061077.

Sáez, C., L.G.M. Ambrosio, S.M. Miguel, J.V. Valcárcel, M.J. Díez, B. Picó, and C. López. 2021. Resistant sources and genetic control of resistance to ToLCNDV in cucumber. *Microorganisms* 9:913. https://doi.org/10.3390/micro organisms9050913.

Sáez, C., C. Esteras, C. Martínez, M. Ferriol, N.P.S. Dhillon, C. López, and B. Picó. 2017. Resistance to tomato leaf curl New Delhi Virus in melon is controlled by a major QTL located in chromosome 11. *Plant Cell Rep.* 36:1571–1584. https://doi.org/10.1007/s00299-017-2175-3.

Sáez, C., C. Martínez, M. Ferriol, S. Manzano, L. Velasco, M. Jamilena, C. López, and B. Picó. 2016. Resistance to tomato leaf curl New Delhi virus in *Cucurbita* spp. *Ann. Appl. Biol.* 169:91–105. https://doi.org/10.1111/aab.12283.

Sáez, C., C. Martínez, J. Montero-Pau, C. Esteras, A. Sifres, J. Blanca, M. Ferriol, C. López, and B. Picó. 2020. A major QTL located in chromosome 8 of *Cucurbita moschata* Is responsible for resistance to tomato leaf curl New Delhi virus. *Front. Plant Sci.* 11:207. https://doi.org/10.3389/fpls.2020.00207. PMID: 32265946; PMCID: PMC7100279.

Sarria-Villada, E., E.G. González, A.I. López-Sesé, A.F. Castiel, and M.L. Gómez-Guillamón. 2009. Hypersensitive response to Aphis gossypii Glover in melon genotypes carrying the Vat gene. *J. Exp. Bot.* 60:3269–3277. https://doi.org/10.1093/jxb/erp163.

Scully, B.T. and W.T. Federer. 1993. Application of genetic theory in breeding for multiple virus resistance. In: M.M. Kyle (ed.). *Resistance to Viral Diseases of Vegetables*. Timber Press, Portland, Ore, p. 167–195.

Sinclair, J.W. 2003. *Screening for resistance to cucurbit yellow stunting disorder virus, gummy stem blight, and monosporascus root rot and detection of RAPD markers associated with QTL for soluble solids, sugars, and vitamin C in melon (Cucumis melo L.)*. PhD Diss., Department of Horticultural Science, Texas A&M University, College Station.

Siskos, L., L. Cui, C. Wang, R.G.F. Visser, Y. Bai, and H.J. Schouten. 2022. A new challenge in melon resistance breeding: The ToLCNDV case. *Euphytica* 218:129. https://doi.org/10.1007/s10681-022-03081-1.

Soria, C., E. Moriones, A. Fereres, E. Garzo, and M.L. Gomez Guillamon. 2003. New source of resistance to mosaic virus transmission by *Aphis gossypii* in melon. *Euphytica* 133 313–318.

Sufrin-Ringwald, T., and Lapidot, M. 2011. Characterization of a synergistic interaction between two cucurbit-infecting begomoviruses; *squash leaf curl virus and watermelon chlorotic stunt virus*. Phytopathol. 101 281–289.

Sugiyama, M. 2013. The present status of breeding and germplasm collection for resistance to viral diseases of cucurbits in Japan. *J. Japanese Soc. Hort. Sci.* 82:193–202. https://doi.org/10.2503/jjshs1.82.193.

Sugiyama, M., Y. Kawazu, N. Fukino, Y. Yoshioka, K. Shimomura, Y. Sakata, and M. Okuda. 2015. Mapping of quantitative trait loci for Melon yellow spot virus resistance in cucumber (*Cucumis sativus* L.). *Euphytica* 205:615–625. https://doi.org/10.1007/s10681-015-1444-x.

Takada, K., K. Kanazawa, K. Takatuka, and T. Kameno. 1979. Breeding of melons resistant to cucumber mosaic virus I. difference in resistance among melon varieties and the regional differences in their distribution. *Bull. Veg. Ornamental Crops Res. Station* 5:1–22.

Tamang, P., K. Ando, W.M. Wintermantel, and J.D. McCreight. 2021. QTL mapping of cucurbit yellow stunting disorder virus resistance in melon accession PI 313970. *HortScience* 56:424–430. https://doi.org/10.21273/HORTSCI15495–20.

Tamang, P., W.M. Wintermantel, and J.D. McCreight. 2020. Suppression of Cucurbit chlorotic yellows virus accumulation in melon breeding line MR-1 under natural infection in Imperial Valley, California. *Cucurbit Genet. Coop. Rpt.* 42:26–29.

Tapley, W.T., W.D. Enzie, and G.P. Van Eseltine. 1937. *The Vegetables of New York. Vol. 1–Part IV. The Cucurbits*. Rpt. N.Y. State Agr. Exp. Sta. for the year ending June 30, 1935. Lyon Co., Albany, NY.

Thomas, C.E. 1985. Resistant reaction of muskmelon line MR-1 against downy mildew. *Phytopathology* 75:504 (Abstr.).

Wang, Z., Q. Gu, H. Sun, H. Li, B. Sun, X. Liang, Y. Yuan, R. Ruili Liu, and Y. Yan Shi. 2014. One-step reverse transcription loop mediated isothermal amplification assay for sensitive and rapid detection of cucurbit chlorotic yellows virus. *J. Virol. Methods* 195:63–66. https://doi.org/10.1016/j.jviromet.2013.08.037.

Webb, R.E. 1963. Local-lesion hosts for some isolates of watermelon mosaic virus. *Plant Dis. Rptr*. 47:1036–1038.

Webb, R.E. 1967. Cantaloupe breeding line B66–5, highly resistant to watermelon mosaic virus 1, *HortScience* 2:58–59.

Whitaker, T.W. and C.I. Jagger. 1937. Breeding and improvement of cucurbits. *In: Yearbook of Agriculture*. U.S. Depaertment of Agriculture, U.S. Government Printing Office, Washington, DC, p. 207–232.

Wintermantel, W.M., L.L.J. Hladky, P. Fashing, K. Ando, and J.D. McCreight. 2019. First report of *cucurbit chlorotic yellows virus infecting* melon in the New World. *Plant Dis.* 103:778. https://doi.org/10.1094/PDIS-08-18-1390-PDN.

Wintermantel, W.M., T. Tian, C. Chen, N. Winarto, S. Szumski, L.J. Hladky, S. Gurung, and J. Palumbo. 2024. Emergence of watermelon chlorotic stunt virus in melon and watermelon in the southwestern United States. *Plant Dis*. https://doi.org/10.1094/PDIS-05-24-1009-PDN. Epub ahead of print. PMID: 39385381.

Wisler, G.C., J.E. Duffus, H.Y. Liu, and R.H. Li. 1998. Ecology and epidemiology of whitefly transmitted closteroviruses. *Plant Dis*. 82:270–280.

Yakoubi, S., C. Desbiez, H. Fakhfakh, C. Wipf-Scheibel, M. Marrakchi, and H. Lecoq. 2007. Occurrence of *Cucurbit yellow stunting disorder virus* and *Cucumber vein yellowing virus* in Tunisia. *J. Plant Pathol*. 89:417–420.

Yousif, M.T., A. Kheyr-Pour, B. Gronenborn, M. Pitrat, and C. Dogimont. 2007. Sources of resistance to Watermelon Chlorotic Stunt Virus in melon. *Plant Breed.* 126:422–427. https://doi.org/10.1111/j.1439-0523.2007.01366.x.

Yuste-Lisbona, F.J., C. Capel, M.L. Gomez-Guillamon, J. Capel, A.I. Lopez-Sese, and R. Lozano. 2011a. Codominant PCR-based markers and candidate genes for powdery mildew resistance in melon (*Cucumis melo* L.). *Theor. Appl. Genet.* 122:747–758.

Yuste-Lisbona, F.J., C. Capel, E. Sarria, R. Torreblanca, M.L. Gomez-Guillamon, J. Capel, R. Lozano, and A.I. Lopez-Sese. 2011b. Genetic linkage map of melon (*Cucumis melo* L.) and localization of a major QTL for powdery mildew resistance. *Mol. Breed.* 27:181–192.

Yuste-Lisbona, F.J., A.I. López-Sesé, and M.L. Gómez-Guillamón. 2009. Inheritance of resistance to races 1, 2 and 5 of powdery mildew in the melon TGR−1551. *Plant Breed.* 129:72–75. https://doi.org/10.1111/j.1439-0523.2009.01655.x.

Zeng, R., F.M. Dai, W.J. Chen, and J.P. Lu. 2011. First report of cucurbit chlorotic yellows virus infecting melon in China. *Plant Dis.* 95:354.

6 Breeding for Fungal Pathogens in Melon (*Cucumis melo* L.)

Venkata Rao Ganaparthi

6.1 FUNGAL DISEASES IN MELON PRODUCTION

The total melon produced around the world was 27.5 million in 2019 (FAO, 2019). The largest producer of melon in the world is China, which produced 13.48 million tons of fruit in 2019 (FAO, 2019). The United States ranks seventh in cantaloupe production. The annual U.S. farm–gate value of melon (*Cucumis melo*) was $325 million in 2022 (USDA ERS, 2023), while watermelon soared to a historical high of $748 million. The annual value of melon exceeded watermelon in the United States from at least 1970 until 2005. Subsequently, the domestic consumption of watermelon has remained relatively constant, while melon has reached a historic low. The USDA Economic Research Service reports per capita availability as an approximation of domestic consumption. Domestic consumption and the corresponding production of melon have decreased by 50% over the past 20 years, falling from a high of 5.17 kg per person and 1150 tons, respectively, to 2.58 kg per person and 550 tons (USDA ERS, 2023). There are currently only four states with reportable production values (California, Arizona, Florida and Georgia), with over 90% in California and Arizona. The major fungal diseases in melon production are powdery mildew, downy mildew, Fusarium wilt, gummy stem blight (GSB) and anthracnose leaf spot. This work provides a comprehensive review of the fungal pathogen causing disease, disease symptoms and genetic nature of the identified resistance.

6.2 POWDERY MILDEW

6.2.1 Pathogen, Symptoms and Pathogenesis

Powdery mildew is caused by biotrophic fungi that infect leaves, stems, flowers and fruits. *Podosphaera xanthii*, *Leveillula taurica* and *Golovinomyces orontii* are the three species reported to cause powdery mildew in Cucurbits (Glawe, 2008; McCreight, 2006). Powdery mildew caused by *P. xanthii* was first reported on melon in the 1920s in California. Although *G. orontii* was the most common causal agent of powdery mildew on melon, after the 1950s, the most common pathogen associated with powdery mildew on melon in the United States is *P. xanthii*. Studies report the evolution of the pathogens as the major reason for the shift in predominance between

DOI: 10.1201/9781003715757-6

the two powdery mildew–causing fungal species. However, the misidentification of the two pathogens is not refuted. The two species did not differ in host range, but the conditions required for the rapid spread of the two pathogens varied. Relatively high temperatures and humidity are optimal for the growth and spread of *P. xanthii*, while *G. orontii* requires lower temperatures to initiate and spread the disease. Powdery mildew caused by *P. xanthii* is a major foliar disease that limits cucurbit production in both the greenhouse and the field (Keinath and DuBose, 2012). Although powdery mildew is a foliar disease, *P. xanthii* is known to infect the hypocotyl, stem, petiole and fruits of cucurbits (Keinath and Dubose, 2004, 2012). Powdery growth on the leaf's upper surface reduces the plant's photosynthetic ability and reduces yield by reducing the size and quality of the fruit (Pérez-García et al., 2009).

Several races of *P. xanthii* have been described based on their virulence on known resistant cultivars (McCreight, 2006). Isolates incapable of overcoming the resistance of the cultivar PMR45, were called race 1; isolates that overcome the resistance of the PMR45 were named race 2. PMR5 is resistant to both race 1 and race 2. Isolates capable of thriving on PMR5 were named race 3 (Hosoya et al., 1999; McCreight and Coffey, 2011). Later, 18 newly emerged races were identified with various race differentials (McCreight and Coffey, 2011). Although several races were identified, races 1 and 2 are the most common causal agents of melon powdery mildew in the United States (McCreight, 2006).

Moderate temperature and high humidity seen in the middle of the fall season are conducive for powdery mildew growth on melon. Usually, in the late spring, conidia are formed from the overwintering *P. xanthii* fruiting structure called a cleistothecium. Ascospores formed on ascii or conidium land on the plant surface. A germ tube is formed after conidial germination on plant surfaces, such as cotyledons, leaves, stems and fruits. Fungus enters the host epidermal cells with the turgor pressure of the appressorium. The fungus forms haustoria upon entering the host cells. The hyphae developing on the plant surfaces appear as white powdery spots, inhibiting the photosynthetic ability and reducing yield and quality of the melons produced. Under high disease pressure, powdery mildew can lead to the death of the plants, resulting in complete yield loss.

The two most common and effective control measures against powdery mildew are resistant cultivars and the periodic application of fungicides. The resistance of *P. xanthii* to fungicides has been reported as early as 1967 when isolates resistant to systemic fungicide 'benomyl' were identified (Schroeder and Provvidenti, 1969). However, a wide array of fungicides with active ingredients such as chlorothalonil, cyflufenamid, flutianil, metrafenone and myclobutanil provides good control against powdery mildew in cucurbits (Vegetable Crop Handbook, 2024). Although fungicides reduce economic losses, they increase the cost of cultivation. Hence, planting cultivars/hybrids that are resistant to different races is an economical and environmentally safe approach to prevent economic losses.

6.2.2 Powdery Mildew–Resistant Sources and Genetics of Resistance

Melon cultivars, breeding lines, landraces and wild relatives have been screened with different races of powdery mildew to identify resistant sources (Hosoya et al., 2000;

María Alvarez et al., 2005; McCreight, 2006; McCreight and Coffey, 2011; Nunes et al., 2017). Powdery mildew–resistant sources were identified, and resistant quantitative trait loci (QTLs) have been discovered in different resistant sources (Branham et al., 2021; López-Martín et al., 2022). Among the diverse germplasm of melon, *C. melon cantalupensis* inbred line 'PMR 5' showed resistance to most races of powdery mildew, including the races standard in the United States, such as race 1, race 2U.S, race 5 and race 6 (Cui et al., 2022). 'PMR5' and 'Edisto 47' together provide resistance for most races of powdery mildew. Surprisingly, no QTL mapping studies have been published using 'PMR 5' as a resistant parent; however, race 1 resistance of 'Edisto 47' was mapped with simple sequence repeat (SSR) markers. Similarly, mapping populations using MR-1 (resistant to race 1, race 2F, race 2U.S., race 4) as resistant parents were employed in identifying powdery mildew race 1–resistant QTLs (Branham et al., 2020; Li et al., 2017). Li et al. (2017) used a mapping population between MR-1 and a susceptible parent 'Top Mark' to identify *P. xanthii* resistant QTLs. Bulk segregant analysis (BSA-seq) and further QTL mapping narrow the QTL region. A single QTL on chromosome 12 was identified, and the QTL region was narrowed to 77.97 kb with 25 SSR markers. Branham et al. (2020) identified two major resistant QTLs, one each on chromosome 5 and chromosome 12, for *P. xanthii* race 1 on cotyledons and true leaves, with a mapping population developed between MR-1 and a susceptible Israeli cultivar 'Ananas Yokneam' (AY). The interaction between these two QTLs was highly significant. However, the nature of the interaction between these QTLs was not established. Mapping studies with 'Edisto 47' as a resistant source of race 1 identified two major dominant QTLs, one each on chromosome 2 and 5 (Ning et al., 2014). Fukino et al. (2008) identified two major QTLs, one each on chromosome 2 and chromosome 12, for two strains of race 1 with a RIL population developed using 'AR5' as a resistant source. TGR-1551 resistance to race 1, race 2F and race 5 was mapped to a major QTL on chromosome 5, explaining at least 50% of phenotypic variance to all three races (Yuste-Lisbona et al., 2011). A single minor resistant QTL on chromosome 8 was identified for the three races. Resistance to all three races was mapped to the same genomic region of chromosome 5 and chromosome 8. Two QTLs, one dominant and one recessive, controlled resistance of TGR-1551 to three *P. xanthii* races, race 1, race 2F and race5, indicating epistasis. However, the type of epistasis was not fully understood as two QTLs are in linkage disequilibrium. Overall, many studies repeatedly reported QTLs on chromosome 2, chromosome 5 and chromosome 12, irrespective of resistant source and race (Yuste-Lisbona et al., 2011; Fukino et al., 2008; Ning et al., 2014; Branham et al., 2021; Li et al., 2017). QTLs identified on these chromosomes are reported as dominant or recessive and show interaction. Dominant and recessively inherited QTLs were identified with resistant sources in *C. melo* subsp. *melon*. While recessively inherited powdery mildew–resistance QTLs have been identified in melon wild species *C. chate* accession C18 and *C. melo var. acidulus* accession PI 313970 (Hong et al., 2015; Mccreight and Coffey, 2011). Although the nature of inheritance was understood for resistant QTLs, epistasis was not fully understood. Further studies are needed to understand the nature of interaction between the discovered QTLs.

Bluemaria graminis f. sp. hordei causes powdery mildew in barley. *B. graminis f. sp. hordei*-barley is the most studied powdery mildew pathogen-plant interaction model (Kusch and Panstruga, 2017). A gene complex 'MLO' plays a crucial role in determining the susceptibility or resistance nature of powdery mildew hosts. Thirteen MLO-like genes were identified in the *C. melo* genome (Zhang et al., 2023). However, none were located on chromosome 5 or chromosome 2, on which many studies reported major powdery mildew–resistant QTLs. Also, the expression of MLO genes was inconsistent with resistant and susceptible bulks. Putative genes found in the QTL interval encode for TIR-NBS-LRR and receptor-like kinase proteins (Branham et al., 2021). In some QTL mapping experiments, genes identified in the QTL interval encode cell wall-fortifying proteins involved in pathogen-triggered immunity (Li et al., 2017). Two candidate genes code for proteins with nucleotide binding sites and leucine-rich repeats, RGA2 and RGA4, were identified in the resistance QTL region of Edisto 47 (Ning et al., 2014). RGA4, together with RGA5, is involved in the identification of effectors and initiation of cell death. So the mechanism of melon resistance to powdery mildew appears to be different from the resistance mechanism of barley.

McCreight (2003) reported distinct genetic controls for resistance to different tissue types. Perchepied et al. (2005b) found tissue-specific-resistant QTLs for cotyledons and true leaves. On the contrary, Branham et al. (2021) found the same major resistant QTLs for powdery mildew across tissues of cotyledons, true leaves and stem. Resistant parent used in the former study is PI 124112, and resistant source for the later study was MR–1. Fukino et al. (2008) identified two resistant QTLs, one each on chromosomes 2 and 12 from the resistant parent 'AR5', irrespective of the tissue type. These observations suggest that tissue-specific genetic control seems to vary by the resistant source.

6.3 DOWNY MILDEW

6.3.1 Pathogen and Symptoms

Cucurbit downy mildew (CDM) caused by *Pseudoperonospora cubensis* is one of the major foliar diseases in cucurbits. Among the cucurbits, cucumber and cantaloupe are more susceptible to downy mildew. CDM is seen on melon in the spring and fall. However, the disease is more destructive on fall-planted melons as prolonged leaf wetness and high relative humidity are common during fall (Keinath, 2019). Most melon hybrids grown are susceptible to *P. cubensis*, hence a serious problem in warmer regions with high rainfall where melon cultivation occurs throughout the year (Savory et al., 2011). In the United States, CDM is typically seen in June or July but has also been reported in April (Keinath et al., 2007). CDM infections dramatically reduce fruit quality by reducing melons' sugar content and reducing yields by up to 80% under high disease pressure (Keinath et al., 2007).

6.3.2 Survival, Multiplication and Spread of the Pathogen

Psudopernospora cubensis produces zoospores within the sporangia on a host plant. Sporangia can be dispersed onto neighboring plants by rain splash and longer distances through wind currents and infected seeds. Under suitable conditions, zoospores germinate (Savory et al., 2011) and the penetration hypha enters the leaf tissue through stomata and forms intercellular hyphae. Infected leaves show chlorosis on the adaxial side of the leaves and white powdery spores on the underside. Soon, chlorotic lesions turn into angular necrotic spots. Under high disease pressure, necrosis expands to the whole leaves and kills the plant. *P. cubensis* was also found in the fruits, seeds, seed cavities and embryos (Cohen et al., 2014). Seedlings grown from the infected seeds showed downy mildew symptoms, proving the seed transmission of downy mildew pathogen.

6.3.3 Pathogen Diversity and Host Differentials

Psudoperanospora cubensis isolates collected from Japan, Israel and the United States were inoculated on 13 Cucurbitaceae hosts from the genera *Benincasa, Citrullus, Cucumis, Cucurbita, Lagenaria, Luffa* and *Mormordica* (Thomas et al., 1987). Based on the differential reaction of isolates on the selected host, *P. cubensis* isolates were classified into five distinct pathotypes. However, all five pathotypes were able to cause disease on *Cucumis melo.* By expanding host differentials, 67 pathotypes were identified worldwide (Cohen, 2015; Labeda et al., 2003; Perchepied et al., 2005a). Although 67 pathotypes were recognized worldwide, only 10 identified in Japan, Israel, the United States, Russia and India are the most commonly causal agents of CDM in melon (Cohen, 2015). Even pathotyping for most seen isolates is time-consuming and requires controlled inoculations of each isolate on host differential. Based on phylogeny, Runge et al. (2011) grouped isolates of *P. cubensis* into two phylogenetic clades. Clade 1 isolates were more genetically diverse in North America, while clade 2 isolates were more genetically diverse in Eurasia, suggesting their respective origins. Clades are host-specific: Clade 1 primarily infects *Cucurbita pepo, Cucurbita maxima, Cucurbita moschata* and *Citrullus lanatus*, whereas hosts of clade 2 isolates are *Cucumis sativus, Cucumis melo* and *Lagenaria siceraria* (Wallace et al., 2020). Along with genetic diversity and host specificity, clade 2 isolates were found earlier in the growing season than clade 1 isolates, indicating a temporal delay between the clade dispersal (Rahman et al., 2021). *P. cubensis* isolates are heterothallic and were further classified into two mating types: A1 and A2 (Cohen et al., 2011).

6.3.4 Management of CDM

Preventive fungicide application earlier in the growing season before the appearance of downy mildew symptoms, followed by a weekly rotation of recommended fungicides with different active ingredients after inoculum arrives, prevents economic loss (Keinath et al., 2019). Fungicides with fluopicolide, carbamates and quinone inside inhibitors are most effective for managing CDM on melon (Ojiambo et al., 2015). Also, sprayings with chlorothalonil, mancozeb and strobilurin controls downy mildew (Keinath et al., 2007). Melon hybrids or cultivars with downy mildew resistance are not widely available, and growers mostly depend on fungicides. Due to the heavy reliance on fungicides, downy mildew isolates resistant to fungicides have been reported (Reuveni et al., 1980). Downy mildew resistance to metalaxyl was reported in Israel, Greece, Italy, Australia and the Czech Republic (Urban and Lebeda, 2006). Another example of fungicide resistance is resistance to strobilurin in Japan (Ishii et al., 2007). As CDM pathogens reproduce sexually (Cohen et al., 2011), new recombination allows rapid evolution of isolates resistant to fungicides. Along with fungicide resistance, isolates overcoming the existing host resistance were seen among the Cucurbits (Holmes and Thomas, 2006). The classic example of host resistance breakdown in cucurbits is new isolates causing the CDM epidemic across the eastern United States on previously resistant cucumbers. Cucumber growers heavily relied on hybrids/cultivars imparted by a single gene: vertical resistance (Olczak-Woltman et al., 2011). As *P. cubensis* reproduces sexually, leading to the rapid evolution of isolates, resistance is imparted by multiple loci (genes), with each locus having a small effect: Horizontal resistance is durable. Hence, cultivar/hybrid with horizontal resistance is the best way to manage CDM on melon.

6.3.5 Resistant Sources and CDM-Resistance Mapping

Screening of available melon germplasm for downy mildew resistance led to identifying a few resistant sources (Thomas, 1992; dos Santos et al., 2009). Thomas (1999) used MR-1, an inbred line obtained from PI 124111, as a resistant source and inferred two incompletely dominant genes, designated as *Pc 1* and *Pc 2*, control CDM resistance of MR-1. Perchepied et al. (2005a) used PI 124112 as a resistant source and tested mapping the population with different isolates of *P. cubensis* in France by both natural and artificial inoculations. A recombinant inbred line (RIL) population developed between PI 124112 and susceptible parent 'Vedrantis' identified one major QTL on linkage group 12, explaining 12%–38% of the phenotypic variance and eight other minor QTLs. Also, a positive linkage between downy mildew and powdery mildew resistance was reported. Toporek et al. (2021, 2023) mapped the resistance MR-1 for two different mating types, A1 and A2. Mating types A1 and A2 were found in equal amounts in southern states, while mating type A1 was most frequent in the Midwest. The RIL population between MR-1 and AY identified two major resistant QTLs for isolates in two different mating types. A stable major resistant QTL at the beginning of chromosome 10 explained 18.7% of phenotypic variation for mating type A2, clade 1. An MLO-like protein-coding gene was identified in the logarithm of odds (LOD) interval of major QTL. An additional eight minor QTLs were identified on different chromosomes. Resistance to isolates in mating type A1 of clade 2 was also mapped onto the genetic map of the MR-1 X AY RIL population. Four QTLs, one on chromosome 5, one on chromosome 8 and two on chromosome 10, were identified. Two QTLs, each on chromosome 8 and chromosome 10 are stable across tests, explaining 22% and 8.3% of phenotypic variance, respectively. Although the major resistant QTLs for isolates in different mating types appears to be different, the correlation between disease scoring of RILs for the two mating types is high. Also, the stable major resistant QTL identified for mating type A2 of clade 1 is 1.2 Mb away from the major stable resistant QTL for mating type A1 of clade 2. Effectors produced by isolates of different clades to induce disease were not shown to be different. Thus, the major CDM-resistant QTL of MR-1 appears to impart broad resistance to different isolates from different mating types. Genes coding for MLO-like proteins (susceptible) and protein kinases (resistant) were found in the QTL intervals. Thus, the downy mildew resistance of MR-1 is broad horizontally and provides resistance to isolates of different mating types.

Kompetitive allele-specific PCR (KASP) markers are developed to track resistance of MR-1 to the isolates of the dual clade (Toporek et al., 2023). However, developed KASP markers need validation prior to use in breeding programs. A proven strategy to develop durable resistant cultivars against pathogens where both susceptible and resistant host genes were identified is to select for parents lacking susceptible genes and with broad resistant genes to breed for CDM resistance (Faris et al., 2013). CDM resistance of MR-1 is co-dominantly inherited (Toporek et al., 2023). As melon hybrids are commercially grown, two CDM-resistant parents are needed to develop hybrids resistant to CDM. ‘Edisto 47’ and MR-1 are the available inbred lines with resistance to different isolates of CDM. ‘Edisto 47’ is an inbred line that produces fruits with salmon-orange flesh color and weighs about 4–5 pounds. Along with downy mildew resistance, ‘Edisto 47’ is highly resistant to different races of powdery mildew and Alternaria leaf spot. Although MR-1 is resistant to different melon pathogens, highly undesirable horticulture traits, such as mealy white fruit flesh and fruit cracking, make it an undesirable parent in a breeding program (Thomas, 1986).

6.4 FUSARIUM WILT

6.4.1 Disease Symptoms and Management

Fusarium oxysporum f. sp. melonis (*Fom*) causes Fusarium wilt in melon. Melon Fusarium wilt results in the death of vines, stunted plants, immature and unmarketable fruits and, in some cases, complete crop loss (Zuniga et al., 1997). Usually, plants express disease symptoms after the fruit set. Initial symptoms are yellowing and wilting of one or a few runners on one side of the plant. As the disease progresses, lesions appear on the runners and visibly extend from the crown to the yellowed tips. In some cases, lesions may appear on the roots along with the red gumming at or below the soil surface. Soon, other runners show similar symptoms, leading to the death of the whole plant. Under high disease pressure, no marketable fruits are harvested, leading to complete crop loss. Like other root pathogens, *Fusarium* enters the Cucurbitaceae host through the root tissue, producing microconidia inside xylem vessels. Microconidia inside vascular bundles germinate and spread pathogens through the xylem vessels of susceptible plants. During the off-season, the pathogen produces thick-walled chlamydospores that can survive up to 10 years in the soil. Chlamydospores are dispersed through water and farm equipment.

Soil fumigation was a traditional and effective measure to control soil-borne pathogens in cucurbits. Nevertheless, an effective fumigant against Fusarium wilt, methyl bromide, has been phased out of the market due to a mandate by the Montreal Protocol and Clean Air Act in 1998. Grafting onto a non-host or resistant rootstock is an effective management practice in cucurbits; however, it increases the cost of cultivation (Keinath and Hassell, 2014). Management practices such as soil solarization and cover cropping with *Vicia villosa*, *Trifolium incarnatum*, *Secale cereale* and

Brassica juncea partly lowered the disease pressure but could not prevent economic losses (Yao et al., 2016; Himmelstein et al., 2014). Thus, growing resistant hybrids/cultivars is the best practice to manage melon Fusarium wilt. Resistance to the most common races of *Fom* is monogenic, and rapid development of cultivars or hybrids with resistance is possible (Martyn, 2014). Since the release of melon hybrids with resistance, Fusarium wilt epidemics have rarely been seen.

6.4.2 Genetics of Fusarium Wilt Resistance in Melon

Four races of *Fusarium oxysporum f. sp. melonis* cause Fusarium wilt *in melon*: *Fom* race 0, *Fom* race 1, *Fom* race 2 and *Fom* race1,2 (Risser et al., 1976; Zink and Thomas, 1990; Zink, 1992). Furthermore, race 1,2 is separated into two races, 1,2y and 1,2w. Among the different races, *Fom* race 1 and *Fom* race 2 are the most common in the melon-growing regions (Zuniga et al., 1997). Screening for resistant sources identified several accessions with resistance to the two most common races of *Fusarium* wilt in melon. MR-1 is resistant to *Fom* race 1 and race 2. Inheritance studies with MR-1 as a resistant parent showed that two single genes impart resistance to *Fom* race 1 and *Fom* race 2 (Risser et al., 1976; Zink, 1992; Zink and Thomas, 1990). Resistance to *Fom* race 0 is co-inherited with *Fom* Race 1 and *Fom* race 2 resistance.

The avirulence gene of *Fom* race 1 is recognized by the host dominant gene *Fom-2*. Randomly amplified polymorphic DNA (RAPD) markers spanning 1.6 kb of *Fom-2* were identified with bulk segregant analysis of families developed between MR-1 and a susceptible Israeli cultivar AY (Wechter et al., 1995). RAPD marker predicted the *Fom-2* gene with 98% accuracy. Later studies developed two user-friendly sequence characterized amplified region (SCAR) markers, MUSKFOM I and MUSKFOM II, to amplify the 1.6-kb region with the *Fom-2* gene, facilitating *Fom* race 1 resistance gene introgression in commercial breeding programs (Wechter et al., 1998). *Fom* race 1 resistance was mapped onto a linkage map with a test-cross population developed between MR-1 and AY (Wang et al., 2000). Two co-dominant amplified fragment length polymorphic (AFLP) markers flanking the *Fom-2* gene at 1.7 and 3.3 cM were identified. AFLP markers were utilized to develop SSRs and sequence tagged sites (STSs) flanking the *Fom-2* (Joobeur et al., 2004). No recombination was found between developed markers (SSR and STS) and the *Fom-2* gene in a population developed between resistant accession 'PI 161375' and 'Vedrantis'. Further SCAR and cleaved amplified polymorphic sequence (CAPS) markers with no recombination with *Fom-2* were developed and utilized in breeding programs (Oumouloud et al., 2015; Wang et al., 2011; Sousaraei et al., 2018) However, SSRs, STS, SCAR and CAPS markers are not high-throughput. Recent studies have mapped *Fom* race 1 resistance onto a high-density linkage map with the bi-parental population developed between MR-1 and AY (Branham et al., 2018). A single QTL on chromosome 11 explained 90% of phenotypic variance. The protein-coding sequence of the candidate gene identified within the QTL interval showed 99% similarity with the *Fom-2* protein. Developed high-throughput KASP markers flanking the major QTL perfectly co-inherited with *Fom* race 1 resistance and accelerate *Fom-2* introgression into melon breeding programs.

Screening of melon bacterand chromosome walking led to the identification and cloning of *Fom-2* (Joobeur et al., 2004). Nearly 1300 gametes were screened with polymerase chain reaction (PCR)–based markers to identify the *Fom* race 1 resistant candidate gene in the recombination cold spot of the *Fom-2* gene interval. Typically, resistance genes are in clusters as an array of paralogues (Hulbert et al., 2001). Unlike typical R genes, no in-paralogs of *Fom-2* were found among the resistance sources and breeding lines or cultivars with resistance to *Fom* race 1 (Joobeur et al., 2004). However, further studies showed the existence of a few alleles of the *Fom-2* gene (Oumouloud et al., 2015). The *Fom-2* gene belongs to the CC subfamily of the nucleotide-binding-site (NBS) leucine-rich-repeat (LRR) family. Usually, R genes of this subfamily have a CC motif before the P-loop. But CC motif of *Fom-2* is in the N-terminal.

Fusarium wilt resistance gene *Fom-2* was overcome by *Fom* race 2 (Risser et al., 1976). Melon line 'Doublon' differentiates race 1 and race 2 as it is susceptible to race 1 and resistant to race 2. Later, screenings with the pathogen identified 'Dulce' and 'Hemed' as resistant to *Fom* race 2. Doublon's, Dulce's and Hemed's resistance is controlled by a single gene, '*Fom-1*' (Danin-Poleg et al., 1999). Later, the *Fom-1* gene was mapped to the linkage group IX (Brotman et al., 2005). Two RFLP markers closely linked to *Fom-1* and homologous to the NBS-LRR genes were identified. However, these two marker haplotypes failed to predict *Fusarium* wilt race 2 resistance among different classes of melon. (Oumouloud et al., 2015, 2010; Pitrat, 2016). Though the authors argue the existence of different alleles of the *Fom-1* gene, the possibility of markers not being in complete linkage disequilibrium (LD) among different classes of cantaloupes is irrefutable. Using resistance gene homologs (RGHs), melon RGHs linked to *Fom-1* were identified on the melon BAC 31016 (Garcia-Mas et al., 2001; Leeuwen et al., 2005). Later, two CAPS markers at 0.4 and 1.2 cM from *Fom-1* were identified (Tezuka et al., 2011). These two markers were validated with inbred lines and commercial hybrids, facilitating MAS in commercial breeding programs. *Fom-1* was further mapped to a 40 kb interval with seven candidate genes. However, none of them belong to the typical NBS-LRR R gene family. Screening of a large backcross population (1190 backcross families) facilitated positional cloning (Brotman et al., 2013). Complete sequence assembly of *Fom-1* was shown to encode a TIR-NB-LRR protein. Interestingly, sequences of *Fom-1* and *Fom-2* are only 13% identical, indicating the specificity of two genes suggesting either independent evolution of the two Fusarium wilt resistance genes of melon or rapid evolution of one or both genes. While no existing evidence exists to support the independent evolution of *Fom-1* and *Fom-2*, the rapid evolution of *Fom-1* is possible as the *Fom-1* gene is in an R gene–rich region with transposable elements in the intron and promoter regions. Transposable elements are known to promote R-gene diversification by inactivating their alleles or by reshuffling the domains. Vaughn et al. (2022) found non-parental haplotypes within the *Fom-1* gene region in the MR-1 × AY bi-parental mapping population, further indicating the rapid evolution of *Fom-1*. cDNA sequences comparison of two susceptible (WMR29 and PI 414723) and three *Fom* race 2 resistant genotypes (Dulce, Vedrantis and MR-1) did not identify any non-synonymous variants but a single synonymous substitution, indicating the possible role of protein folding and binding with the pathogen secreted protein in imparting resistance to

Fom race 2. Later, *Fom* race 2 resistance was mapped onto a high-density bi-parental linkage map. Two neighboring QTLs on chromosome 9 explain 40% and 35% of phenotypic variance to *Fom* race 2 resistances of MR-1, respectively. Due to a gene duplication event, two identified QTLs includes an identical copy of the previously identified NBS-LRR gene (Vaughn et al., 2022). KASP markers were developed for the SNP variants within the two resistant QTL regions and were validated with the two segregating populations developed from resistant RIL-206 and susceptible 'Top Mark' and 'Charentais'. A single haplotype with four KASP markers tracked *Fom* race 2 resistance with 91% accuracy.

A more virulent race able to overcome the resistance of both *Fom-1* and *Fom-2* genes was named race 1.2 (Risser et al., 1976). However, the inheritance of resistance to race 1.2 was partially polygenic, with two complementary recessive genes providing resistance to race 1.2 (Herman and Perl-Treves, 2007; Perchepied et al., 2005b). Oumouloud et al. (2013) noticed isolate-specific resistance among different cultivars. Further studies are needed to understand the resistance to *Fom* race 1.2 among the melon accessions and cultivars.

Several *Fusarium* species were shown to cause fruit rot of melon and is an economically significant post-harvest disease reported in Thailand, China and Brazil (Li et al., 2019; Medeiros Araújo et al., 2021). *Fusarium falciforme*, *F. sulawesiense, F. pernambucanum* and *F. kalimantanense* cause melon fruit rot, with *F. falicforme* being the most common and most aggressive species in Brazil (Medeiros Araújo et al., 2021). *Fusarium falciforme* and *Fusarium annulatum* were shown to cause fruit rot in Spain. In China, *Fusarium equiseti* causes melon fruit rot. *Fusarium* fruit rot directly affects consumers as symptoms appear only after the harvest. According to social media, reduced melon consumption in the United States is correlated with melon associated human pathogen outbreaks. Melon-associated salmonella outbreaks combined with fruit rot could decimate the melon market. Fruit rot phenotyping tests need to be developed for screening germplasm and breeding resistant hybrids.

6.5 GSB

GSB is a major disease seen in cucurbits and is common in most cucurbit-producing areas around the world (Keinath, 2011). Melon is highly susceptible to GSB among the cucurbits. Three genetically distinct species of *Stagonosporopsis* cause GSB in cucurbits (Brewer et al., 2015); Stewart et al. (2015): *Stagnosporopsis citrulli, Stagnosporopsis cucurbitacearum* and *Stagnosporopsis caricae. Stagnosporopsis citrulli* is the most common species on melon in the southeastern United States, while *Stagnosporopsis cucurbitacearum* is more common in the northeastern and midwestern United States.

6.5.1 Disease Symptoms, Pathogen Survival and Spread and Control Measures

The pathogen can infect all the melon plant parts above ground. Symptoms on cotyledons, leaves, petioles, hypocotyls and stems include circular or triangular tan-to-brown water-soaked lesions. On the leaves, lesions start at the margin or touch the margins and are occasionally found at the mid-vein. The center of the spot is darker than the surrounding portions. Adjacent leaf spots expand and merge with each other, leading to blight appearing on the leaves. Petiole lesions cause the entire leaf to collapse. Under severe disease pressure, along with the leaf symptoms, cankers appear on the crown, the main stem and the vines. Light brown cankers appear as water-soaked green lesions that later become rough and dry. Occasionally, cracks appear on the cankers. Amber plant sap may appear as gum on the cracked cankers. Although the common name GSB indicates 'stem oozing gum', such symptoms are not a diagnostic symptom for GSB, as such symptoms are also associated with physical injuries and Fusarium wilt. The pathogen also causes fruit black rot, which starts in the cracks of the fruit or stem end of the fruit. The fruit flesh becomes watery and discolored, and black mold is seen in the seed cavity. Black rot is more often seen on western-type cantaloupes that produce heavily netted, lightly sutured fruit with firm ripe flesh than eastern-type muskmelons with deep sutures, lightly netted fruits with soft ripe flesh.

During the off-season, the fungus survives in the infected crop debris (Keinath, 2008). The pathogen survives 7–24 months on crop residuals. As favorable conditions appear, the pathogen produces pseudothecia, which are spread by air. Complete removal of crop debris is highly difficult and cropping fall cucurbits in the same field after spring cucurbits leads to severe GSB outbreaks. Crop rotation with non-cucurbits reduces GSB and increases marketable yield (Sitterly, 1969). Fungicide application is necessary to prevent yield losses. Fungicides with active ingredients tebuconazole, prothioconazole, chlorothalonil, cymoxanil, fluopyram and difenoconazole (Vegetable crop handbook for southeastern United States, 2024). GSB isolates resistant to fungicides were first seen in Georgia in the United States in 1999 (Stevenson et al., 2004). Azoxystrobin was used as an effective fungicide against GSB on cucurbits in the 1990s. By 2001 and 2002, 91% of *Didymella bryoniae* isolates found in Georgia were resistant to Azoxystrobin. Tebuconazole is a generic fungicide commonly used to control GSB on cucurbits (SC Grower, 2022). Only 3% of GSB pathogen isolates collected on cucurbits showed sensitivity to tebuconazole.

6.5.2 Resistance Sources and Genetics of GSB Resistance

Due to the importance of GSB in melon production and the rapid evolution of fungicide-resistant isolates, extensive germplasm screening was undertaken to identify resistant sources (Nortan and Krishna, 1967). PI 140471 was the earliest accession used as a GSB-resistant source. Breeding lines with resistance to GSB were developed using PI 140471 as a resistant parent. However, the resistance of developed breeding lines was not enough to prevent economic losses (Zhang et al., 1997). Further resistance screening identified four accessions: PI 157082, PI 511890, PI 482398 and PI 482399, with resistance stronger than PI 140471. The resistance of both PI 157082 and PI 511890 was shown to be controlled by a single dominant gene and was different from the loci identified from PI 140471. Inheritance studies showed the resistance nature of PI 482399 as recessive in nature. In contrast, Frantz and Jahn's (2004) inheritance study showed resistance of PI 482398 as dominant. The resistance locus of PI 140471 was named *Gsb-1*, the resistance of PI 157082 was named *Gsb-2* and the resistance loci of PI 511890 and PI 482398 were named *Gsb-3* and *Gsb-4*, respectively. The recessive resistant locus of PI 482399 was designated as *gsb-5* (Hassan et al., 2018). Later screenings identified PI 420145 as strongly resistant to GSB, and its dominant resistance gene is *Gsb-6* (Wolukau et al. 2007). Ma et al. (2023) identified an accession (H55R) immune to GSB. Bulk-segregate analysis with H55R as a resistant parent identified a single resistant QTL on chromosome 7 and was designated as *Gsb-7(t)*. Different resistant loci do not show synergy in imparting GSB resistance, indicating the possibility of different resistance mechanisms among resistant accessions (Zuniga et al., 1999).

Synteny between the cucumber GSB-resistant region on chromosome 5 and melon resistance on chromosome 9 identified a TIR-NBS-LRR gene (Hassan et al., 2018). A sequence comparison of the identified TIR-NBS-LRR gene between resistant and susceptible lines found deletions, insertions, and frameshift mutations in the introns and exons of the TIR-NBS-LRR gene. InDel markers developed from the sequence comparison of resistant and susceptible lines completely co-segregated with resistance. Based on the BSA of the population segregating for GSB resistance of H55R, the sequence of a single gene, *MELO3C010403*, with two transcripts varied between resistant and susceptible proteins, has five exons and four introns (Ma et al., 2023). The expression of one transcript, *MELO3C010403-T2*, was threefold higher in the resistant parent than the susceptible parent and is most likely the candidate gene. The transcript belongs to a receptor-like kinase family protein with the transmembrane helix region and the cytoplasmic serine/threonine-protein kinase domain at the C-terminus. HS is a GSB-resistant inbreed line (Yang et al., 2020). Two GSB-resistant QTLs of HS were discovered on chromosome 2 and chromosome 5 with a RIL population. A microbe-associated molecular pattern recognizing a transmembrane receptor protein, LRR receptor-like serine/threonine-protein kinase (LRR-RLK), was found as the best candidate gene for GSB resistance of HS. A non-synonymous SNP was identified in the coding sequence of identified LRR-RLK and co-segregated with known resistance and susceptible lines.

6.6 ALTERNARIA LEAF BLIGHT

6.6.1 Disease Symptoms, Spread and Control Measures

Alternaria cucumerina, *Alternaria alternata* and *Alternaria solani* are known to cause alternaria leaf blight (ALB) of cantaloupe around the world (Zhou and Everts, 2008; Ahmed et al., 2021; Zhao et al., 2016; Ntui et al., 2010). ALB is a significant problem in the midwestern and eastern states of the United States, reducing yields by up to 50%. Along with yield losses, fruit quality is significantly affected by severe ALB infection. The disease starts as dark brown spots or lesions; the center of the lesion is light tan, surrounded by a dark brown ring and a chlorotic halo (Zhou and Everts, 2008). ALB starts on the older leaves and progresses to younger leaves. As the disease progresses, leaf splitting is seen on infected older leaves. Disease symptoms are usually seen on the edges of the older leaves, and lesions are not seen on the stems and fruit. Heavy rains accompanied by wind for prolonged periods promote ALB outbreaks. Strategic fungicide application with MelCast controlled ALB on muskmelon (Egel and Harmon, 2001). Along with chemicals against ALB, resistant cultivars/hybrids were released by Cornell University.

Two major factors responsible for the ALB establishment and spread are temperature and leaf wetness duration (Evans et al., 1992). The optimum temperature and leaf wetness duration required for disease establishment were 18 °C and 8 h. ALB is not seen at higher temperatures (27–30 °C) irrespective of leaf wetness duration. Along with leaf wetness duration and temperatures, foliar damage and the number of overwintered mycelia in the field affect disease severities (Jackson and Weber, 1959; Chandler and Thomas, 1991). An increase in foliar damage and overwintered mycelia increases disease severity. Mancozeb- and chlorothalonil-based fungicides are commonly used to control ALB (Keinath et al., 2007). However, resistant cultivars are the most effective management strategy (Boyhan and Norton, 1992).

6.6.2 Resistant Sources and Genetics of ALB Resistance

Genetic studies on ALB resistance are limited. MR-1 is an inbred line resistant to multiple diseases. ALB resistance of MR-1 was shown to be controlled by a single dominant gene and was designated as *Ac* Thomas et al. (1990). AC-82–37 is an inbred line with resistance to ALB (Boyhan and Norton, 1992). Inheritance studies showed the resistance nature of AC-82–37 as additive, dominant and epistatic. The RIL population developed between MR-1 and AY was screened for ALB resistance (Daley et al., 2017). Two QTLs, one each on chromosomes 10 and 12, were identified and explained 33% of phenotypic variance together, indicating the oligogenic nature of MR-1's ALB resistance.

REFERENCES

Ahmed, Muhammad Zeshan, Saba Saeed, Ahmed Hassan, Salman Ghuffar, Ahsan Abdullah, Rashid Iqbal, Hafiz Arslan Anwaar, Khansa Haque, Muhammad Ammad Asif, and Muhammad Subhan Shafique. 2021. "Alternaria Alternata Causing Leaf Spot of Cucumis Melo (Muskmelon) in Pakistan." *The American Phytopathological Society* 105 (6). https://doi.org/10.1094/PDIS-05-20-0973-PDN.

Boyhan, G. and Norton, J. 1992. "Inheritance of Resistance to Alternaria Leaf Blight in Muskmelons." *HortScience* 27 (10): 1114–15.

Branham, Sandra E., Chandrasekar Kousik, Mihir K. Mandal, and W. Patrick Wechter. 2021. "Quantitative Trait Loci Mapping of Resistance to Powdery Mildew Race 1 in a Recombinant Inbred Line Population of Melon." *Plant Disease* 105 (12). American Phytopathological Society. https://doi.org/10.1094/PDIS-12-20-2643-RE/ASSET/IMAGES/LARGE/PDIS-12-20-2643-RET5-1642709760698.JPEG.

Branham, Sandra E., Amnon Levi, Melanie Katawczik, Zhangjun Fei, and W. Patrick Wechter. 2018. "Construction of a Genome-Anchored, High-Density Genetic Map for Melon (*Cucumis melo* L.) and Identification of Fusarium Oxysporum f. Sp. Melonis Race 1 Resistance QTL." *Theoretical and Applied Genetics* 131 (4). Springer Verlag: 829–37. https://doi.org/10.1007/s00122-017-3039-5.

Branham, Sandra E., W. Patrick Wechter, Kai Shu Ling, Bidisha Chanda, Laura Massey, Guangwei Zhao, Nihat Guner, et al. 2020. "QTL Mapping of Resistance to Fusarium Oxysporum f. Sp. Niveum Race 2 and Papaya Ringspot Virus in Citrullus Amarus." *Theoretical and Applied Genetics* 133 (2). Springer: 677–87. https://doi.org/10.1007/s00122-019-03500-3.

Brewer, M., Rath, M., Li, H. 2015. "Genetic Diversity and Population Structure of Cucurbit Gummy Stem Blight Fungi Based on Microsatellite Markers." *Phytopathology* 105 (6): 815–24. https://doi.org/10.1094/PHYTO-10-14-0282-R

Brotman, Yariv, Irina Kovalski, Catherine Dogimont, Michel Pitrat, Vitaly Portnoy, Nurit Katzir, and Rafael Perl-Treves. 2005. "Molecular Markers Linked to Papaya Ring Spot Virus Resistance and Fusarium Race 2 Resistance in Melon." *Theoretical and Applied Genetics* 110 (2): 337–45. https://doi.org/10.1007/s00122-004-1845-z.

Brotman, Yariv, Michael Normantovich, Zachi Goldenberg, Zvi Zvirin, Irina Kovalski, Nastacia Stovbun, Tirza Doniger, et al. 2013. "Dual Resistance of Melon to Fusarium Oxysporum Races 0 and 2 and to Papaya Ring-Spot Virus Is Controlled by a Pair of Head-to-Head-Oriented Nb-Lrr Genes of Unusual Architecture." *Molecular Plant*. Oxford University Press. https://doi.org/10.1093/mp/sss121.

Chandler, L. and Thomas, C. 1991. "Effect of Leaf Miner Feeding-Activity on the Incidence of Alternaria Leaf-Blight Lesions on Muskmelon Leaves." *Plant Disease* 75 (9): 938–40.

Cohen, Y. 2015. "The Novel Oomycide Oxathiapiprolin Inhibits All Stages in the Asexual Life Cycle of Pseudoperonospora Cubensis – Causal Agent of Cucurbit Downy Mildew." *PLoS ONE* 10 (10). Public Library of Science: e0140015. https://doi.org/10.1371/JOURNAL.PONE.0140015.

Cohen, Y., A. E. Rubin, and M. Galperin. 2011. "Formation and Infectivity of Oospores of Pseudoperonospora Cubensis, the Causal Agent of Downy Mildew in Cucurbits." *The American Phytopathological Society* 95 (7): 874. https://doi.org/10.1094/PDIS-02-11-0127.

Cohen, Y., Avia E. Rubin, Mariana Galperin, Sebastian Ploch, Fabian Runge, and Marco Thines. 2014. "Seed Transmission of Pseudoperonospora Cubensis." *PLoS ONE* 9 (10). Public Library of Science: e109766. https://doi.org/10.1371/JOURNAL.PONE.0109766.

Cui, Lei, Lampros Siskos, Chen Wang, Henk J. Schouten, Richard G. F. Visser, and Yuling Bai. 2022. "Breeding Melon (Cucumis Melo) with Resistance to Powdery Mildew and Downy Mildew." *Horticultural Plant Journal* 8 (5). Elsevier: 545–61. https://doi.org/10.1016/J.HPJ.2022.07.006.

Danin-Poleg, Y., Reis, N., Baudracco-Arnas, S., Pitrat, M., Staub, E., Oliver, M., Arus, P., deVicente, C., and Katzir, N. 1999. "Simple Sequence Repeats in Cucumis Mapping and Map Merging." Genome 43: 963–74.

Daley, James, Sandra Branham, Amnon Levi, Richard Hassell, and Patrick Wechter. 2017. "Mapping Resistance to Alternaria Cucumerina in Cucumis Melo." *Phytopathology* 107 (4). *American Phytopathological Society*: 427–32. https://doi.org/10.1094/PHYTO-06-16-0246-R/SUPPL_FILE/PHYTO-06-16-0246-R.SF1.PDF.

dos Santos, Gil R., Castro, M. D., Ramos, L. N., Cafe-Filcho, A. C., Reis, A., Momente, V. G., Peluzio, J. M., and Ignacio, M. 2009. "Reaction of Melon Genotypes to the Gummy Stem Blight and The Downy Mildew." *Horticultura Brasileira* 27: 160–65.

Egel, D. S., and P. Harmon. 2001. "Effects of Nozzle Type and Spray Pressure on Control of Alternaria Leaf Blight of Muskmelon with Chlorothalonil." *Plant Disease* 85 (10). American Phytopathological Society: 1081–84. https://doi.org/10.1094/PDIS.2001.85.10.1081.

Evans, K., Nyquist, W., Latin, R. 1992. "A Model Based on Temperature and Leaf Wetness Duration for Establishment of Alternaria Leaf-Blight of Muskmelon." *Phytopathology* 82 (8): 890–95.

FAO. 2019. Food and Agriculture Organization of the United Nations. Available online: http://www.fao.org/faostat/en/#data/QC (accessed on 10 January 2025)

Faris, Justin D., Zhaohui Liu, and Steven S. Xu. 2013. "Genetics of Tan Spot Resistance in Wheat." *Theoretical and Applied Genetics* 126 (9): 2197–2217. https://doi.org/10.1007/s00122-013-2157-y.

Frantz, J. D., and M. M. Jahn. 2004. "Five Independent Loci Each Control Monogenic Resistance to Gummy Stem Blight in Melon (*Cucumis melo* L.)." *Theoretical and Applied Genetics* 108 (6). Springer: 1033–38. https://doi.org/10.1007/S00122-003-1519-2/TABLES/5.

Fukino, Nobuko, Takayoshi Ohara, Antonio J. Monforte, Mitsuhiro Sugiyama, Yoshiteru Sakata, Miyuki Kunihisa, and Satoru Matsumoto. 2008. "Identification of QTLs for Resistance to Powdery Mildew and SSR Markers Diagnostic for Powdery Mildew Resistance Genes in Melon (*Cucumis melo* L.)." *Theoretical and Applied Genetics* 118 (1). Springer: 165–75. https://doi.org/10.1007/S00122-008-0885-1/TABLES/5.

Garcia-Mas, Jordi, Hans Van Leeuwen, Amparo Monfort, M. Carmen de Vicente, Pere Puigdomènech, and Pere Arús. 2001. "Cloning and Mapping of Resistance Gene Homologues in Melon." *Plant Science* 161 (1). Elsevier: 165–72. https://doi.org/10.1016/S0168-9452(01)00403–4.

Glawe, Dean A. 2008. "Ar-223016 IP: 198.21.215.119 On: Wed, 04 Sept 2024 17:24:30 Downloaded from Www.Annualreviews.Org" 24: 30. https://doi.org/10.1146/annurev.phyto.46.081407.104740.

Hassan, Md. Zahid, Md. Abdur Rahim, Sathishkumar Natarajan, Arif Hasan Khan Robin, Hoy Taek Kim, Jong In Park, and Ill Sup Nou. 2018. "Gummy Stem Blight Resistance in Melon: Inheritance Pattern and Development of Molecular Markers." *International Journal of Molecular Sciences* 19 (10). Multidisciplinary Digital Publishing Institute (MDPI). https://doi.org/10.3390/IJMS19102914.

Herman, R., and R. Perl-Treves. 2007. "Characterization and Inheritance of a New Source of Resistance to Fusarium Oxysporum f. Sp. Melonis Race 1.2 in Cucumis Melo." *The American Phytopathological Society* 91 (9): 1180–86. https://doi.org/10.1094/PDIS-91-9-1180.

Himmelstein, J. C., J. E. Maul, and K. L. Everts. 2014. "Impact of Five Cover Crop Green Manures and Actinovate on Fusarium Wilt of Watermelon." *Plant Disease* 98 (7). American Phytopathological Society: 965–72. https://doi.org/10.1094/PDIS-06-13-0585-RE.

Holmes, G., and Thomas, C. 2006. "The History and Re-emergence of Cucurbit Downy Mildew." *Phytopathology* 99: 171.

Hong, C., K. Weiping, L. Junfen, and L. Jiping. 2015. "Analysis of Powdery Mildew Resistance in Wild Melon MLO Mutants." *Horticultural Plant Journal* 1 (3). Elsevier: 165–71. https://doi.org/10.16420/J.ISSN.2095–9885.2015–0036.

Hosoya, K., M. Kuzuya, T. Murakami, K. Kato, K. Narisawa, and H. Ezura. 2000. "Impact of Resistant Melon Cultivars on Sphaerotheca Fuliginea." *Plant Breeding* 119 (3). John Wiley & Sons, Ltd: 286–88. https://doi.org/10.1046/J.1439-0523.2000.00489.X.

Hosoya, K., K. Narisawa, M. Pitrat, and H. Ezura. 1999. "Race Identification in Powdery Mildew (Sphaerotheca Fuliginea) on Melon (Cucumis Melo) in Japan." *Plant Breeding* 118 (3). John Wiley & Sons, Ltd: 259–62. https://doi.org/10.1046/J.1439-0523.1999.118003259.X.

Hulbert, S. H., C. A. Webb, S. M. Smith, and Q. Sun. 2001. "Resistance Gene Complexes: Evolution and Utilization." *Annual Review of Phytopathology* 39: 285–312. https://doi.org/10.1146/ANNUREV.PHYTO.39.1.285.

Ishii, H., B. A. Fraaije, T. Sugiyama, K. Noguchi, K. Nishimura, T. Takeda, T. Amano, and D. W. Hollomon. 2007. "Occurrence and Molecular Characterization of Strobilurin Resistance in Cucumber Powdery Mildew and Downy Mildew." *The American Phytopathological Society* 91 (12): 1166–71. https://doi.org/10.1094/PHYTO.2001.91.12.1166.

Jackson, C., and Weber, G. 1959. "Morphology and Taxonomy of Alternaria Cucumerina." *Mycologia* 51 (3): 401–08.

Joobeur, Tarek, Joseph J. King, Shelly J. Nolin, Claude E. Thomas, and Ralph A. Dean. 2004. "The Fusarium Wilt Resistance Locus Fom-2 of Melon Contains a Single Resistance Gene with Complex Features." *Plant Journal* 39 (3): 283–97. https://doi.org/10.1111/J.1365-313X.2004.02134.X.

Keinath, Anthony P. 2011. "From Native Plants in Central Europe to Cultivated Crops Worldwide: The Emergence of Didymella bryoniae as a Cucurbit Pathogen." *HortScience* 46 (4): 532–35.

Keinath, Anthony P. 2008. "Survival of Didymella Bryoniae in Infested Muskmelon Crowns in South Carolina." *The American Phytopathological Society* 92 (8): 1223–28. https://doi.org/10.1094/PDIS-92-8-1223.

Keinath, Anthony P. 2019. "Integrated Management of Downy Mildew on Slicing Cucumber with Fungicides and Host Resistance but Not Trellising." *Plant Disease* 103 (10). American Phytopathological Society: 2592–98. https://doi.org/10.1094/PDIS-02-19-0323-RE/ASSET/IMAGES/LARGE/PDIS-02-19-0323-RE_F2-1568029371426.JPEG.

Keinath, Anthony P., and Virginia B. Dubose. 2004. "Evaluation of Fungicides for Prevention and Management of Powdery Mildew on Watermelon." *Crop Protection* 23 (1). Elsevier: 35–42. https://doi.org/10.1016/S0261-2194(03)00165–0.

Keinath, Anthony P., and Virginia B. DuBose. 2012. "Controlling Powdery Mildew on Cucurbit Rootstock Seedlings in the Greenhouse with Fungicides and Biofungicides." *Crop Protection* 42 (December). Elsevier: 338–44. https://doi.org/10.1016/J.CROPRO.2012.06.009.

Keinath, Anthony P., K. L. Everts, D. B. Langston, D. S. Egel, and G. J. Holmes. 2007. "Multi-State Evaluation of Reduced-Risk Fungicides and Melcast against Alternaria Leaf Blight and Gummy Stem Blight on Muskmelon." *Crop Protection* 26 (8). Elsevier: 1251–58. https://doi.org/10.1016/J.CROPRO.2006.10.024.

Keinath, Anthony P., and Richard L. Hassell. 2014. "Suppression of Fusarium Wilt Caused by Fusarium Oxysporum f. Sp. Niveum Race 2 on Grafted Triploid Watermelon." https://doi.org/10.1094/PDIS-01-14-0005-RE.

Keinath, Anthony P., G. J. Holmes, K. L. Everts, D. S. Egel, and D. B. Langston. 2007. "Evaluation of Combinations of Chlorothalonil with Azoxystrobin, Harpin, and Disease Forecasting for Control of Downy Mildew and Gummy Stem Blight on Melon." *Crop Protection* 26 (2). Elsevier: 83–88. https://doi.org/10.1016/J.CROPRO.2006.04.004.

Keinath, Anthony P., Sally A. Miller, and Christine D. Smart. 2019. "Response of Pseudoperonospora Cubensis to Preventative Fungicide Applications Varies by State and Year." *The American Phytopathological Society* 20 (3): 142–46. https://doi.org/10.1094/PHP-04-19-0028-RS.

Kusch, Stefan, and Ralph Panstruga. 2017. "Mlo-Based Resistance: An Apparently Universal 'Weapon' to Defeat Powdery Mildew Disease." *Molecular Plant-Microbe Interactions* 30 (3). American Phytopathological Society: 179–89. https://doi.org/10.1094/MPMI-12-16-0255-CR/ASSET/IMAGES/LARGE/MPMI-12-16-0255-CR_T1-1491248712271.JPEG.

Leeuwen, Hans Van, Jordi Garcia-Mas, María Coca, Pere Puigdoménech, and Amparo Monfort. 2005. "Analysis of the Melon Genome in Regions Encompassing TIR-NBS-LRR Resistance Genes." *Molecular Genetics and Genomics* 273 (3). Springer: 240–51. https://doi.org/10.1007/S00438-004-1104-7/FIGURES/5.

Li, Bing, Yulong Zhao, Qianglong Zhu, Zhipeng Zhang, Chao Fan, Sikandar Amanullah, Peng Gao, and Feishi Luan. 2017. "Mapping of Powdery Mildew Resistance Genes in Melon (*Cucumis melo* L.) by Bulked Segregant Analysis." *Scientia Horticulturae* 220 (June). Elsevier: 160–67. https://doi.org/10.1016/J.SCIENTA.2017.04.001.

Li, Y. G., R. Zhang, L. Meng, E. Ali, P. Ji, Q. F. Zhang, and G. W. Cui. 2019. "Occurrence of Fruit Rot of Cantaloupe Caused by Fusarium Equiseti in China." *Plant Disease* 103 (10). https://doi.org/10.1094/PDIS-03-19-0671-PDN.

López-Martín, María, Ana Pérez-de-Castro, Belén Picó, and María Luisa Gómez-Guillamón. 2022. "Advanced Genetic Studies on Powdery Mildew Resistance in TGR−1551." *International Journal of Molecular Sciences* 23 (20). MDPI: 12553. https://doi.org/10.3390/IJMS232012553/S1.

Ma, Jian, Congcong Li, Jiaxing Tian, Yanhong Qiu, Lihua Geng, and Jianshe Wang. 2023. "Identification and Fine Mapping of Gummy Stem Blight Resistance Gene Gsb-7(t) in

Melon." *Phytopathology* 113 (5). American Phytopathological Society: 858–65. https://doi.org/10.1094/PHYTO-05-22-0169-R/ASSET/IMAGES/LARGE/PHYTO-05-22-0169-RF4-1688027980667.JPEG.

María Alvarez, José, Rafael González-Torres, Cristina Mallor, and María Luisa Gómez-Guillamón. 2005. "Potential Sources of Resistance to Fusarium Wilt and Powdery Mildew in Melons." *Hortscience* 40 (6).

Martyn, Ray D. 2014. "Fusarium Wilt of Watermelon: 120 Years of Research." In *Horticultural Reviews*, Vol. 42, pp. 349–442. Hoboken, New Jersey: John Wiley & Sons, Inc. https://doi.org/10.1002/9781118916827.ch07.

Mccreight, James D. 2003. "Genes for Resistance to Powdery Mildew Races 1 and 2 U.S. in Melon PI 313970." *HortScience* 38:591.

McCreight, James D. 2006. "Melon-Powdery Mildew Interactions Reveal Variation in Melon Cultigens and Podosphaera Xanthii Races 1 and 2." *Journal of the American Society for Horticultural Science* 131.

Mccreight, James D., and Michael D. Coffey. 2011. "Inheritance of Resistance in Melon PI 313970 to Cucurbit Powdery Mildew Incited by Podosphaera Xanthii Race S." *Hortscience* 46.

Medeiros Araújo, Maria B., Gláucia M. Moreira, Luan Vítor Nascimento, Geovane de Almeida Nogueira, Selma Rogéria de C. Nascimento, Ludwig H. Pfenning, and Márcia Michelle de Q. Ambrósio. 2021. "Fusarium Rot of Melon Is Caused by Several Fusarium Species." *Plant Pathology* 70 (3). Blackwell Publishing Ltd: 712–21. https://doi.org/10.1111/PPA.13328.

Ning, Xuefei, Xianlei Wang, Xingwang Gao, Ziqiang Zhang, Lihu Zhang, Weili Yan, and Guan Li. 2014. "Inheritances and Location of Powdery Mildew Resistance Gene in Melon Edisto47." *Euphytica* 195 (3). Kluwer Academic Publishers: 345–53. https://doi.org/10.1007/S10681-013-1000-5/FIGURES/2.

Norton, J., and Krishna P. 1967. "Inheritance of resistance to Mycosphaerella citrullina in muskmelon." *HortScience* 216–17.

Ntui, Valentine Otang, Gunaratnam Thirukkumaran, Pejman Azadi, Raham Sher Khan, Ikuo Nakamura, and Masahiro Mii. 2010. "Stable Integration and Expression of Wasabi Defensin Gene in 'Egusi' Melon (Colocynthis Citrullus L.) Confers Resistance to Fusarium Wilt and Alternaria Leaf Spot." *Plant Cell Reports* 29 (9). Springer: 943–54. https://doi.org/10.1007/S00299-010-0880-2/FIGURES/7.

Nunes, E. W. L. P., C. Esteras, A. O. Ricarte, E. Martínez-Perez, M. L. Gómez-Guillamón, G. H. S. Nunes, and M. B. Picó. 2017. "Brazilian Melon Landraces Resistant to Podosphaera Xanthii Are Unique Germplasm Resources." *Annals of Applied Biology* 171 (2). John Wiley & Sons, Ltd: 214–28. https://doi.org/10.1111/AAB.12370.

Ojiambo, Peter S., David H. Gent, Lina M. Quesada-Ocampo, Mary K. Hausbeck, and Gerald J. Holmes. 2015. "Epidemiology and Population Biology of Pseudoperonospora Cubensis: A Model System for Management of Downy Mildews." *Annual Review of Phytopathology* 53. Annual Reviews Inc.: 223–46. https://doi.org/10.1146/ANNUREV-PHYTO-080614-120048/CITE/REFWORKS.

Olczak-Woltman, Helena, Joanna Marcinkowska, and Katarzyna Niemirowicz-Szczytt. 2011. "The Genetic Basis of Resistance to Downy Mildew in Cucumis Spp. – Latest Developments and Prospects." *Journal of Applied Genetics* 52 (3): 249. https://doi.org/10.1007/S13353-011-0030-8.

Oumouloud, A., M. S. Arnedo-Andrés, R. González-Torres, and J. M. Álvarez. 2010. "Inheritance of Resistance to Fusarium Oxysporum f. Sp. Melonis Races 0 and 2 in Melon Accession Tortuga." *Euphytica* 176 (2): 183–89. https://doi.org/10.1007/s10681-010-0201-4.

Oumouloud, A., M. El Otmani, and J. M. Álvarez. 2015. "Molecular Characterization of Fom-1 Gene and Development of Functional Markers for Molecular Breeding of Resistance to

Fusarium Race 2 in Melon." *Euphytica* 205 (2). Kluwer Academic Publishers: 491–501. https://doi.org/10.1007/s10681-015-1420-5.

Oumouloud, A., R. González Torres, A. Garcés-Claver, H. Chikh-Rouhou, and J. M. Alvarez. 2013. "Differential Response of Cucumis Melo to Fusarium Oxysporum f. Sp. Melonis Race 1.2 Isolates." *Crop Protection* 44 (February). Elsevier: 91–94. https://doi.org/10.1016/J.CROPRO.2012.10.001.

Perchepied, L., M. Bardin, C. Dogimont, and M. Pitrat. 2005a. "Relationship between Loci Conferring Downy Mildew and Powdery Mildew Resistance in Melon Assessed by Quantitative Trait Loci Mapping." *Phytopathology* 95 (5): 556–65. https://doi.org/10.1094/PHYTO-95-0556.

Perchepied, L., C. Dogimont, and M. Pitrat. 2005b. "Strain-Specific and Recessive QTLs Involved in the Control of Partial Resistance to Fusarium Oxysporum f. Sp. Melonis Race 1.2 in a Recombinant Inbred Line Population of Melon." *Theoretical and Applied Genetics* 111 (1). Springer: 65–74. https://doi.org/10.1007/S00122-005-1991-Y/TABLES/4.

Pérez-García, Alejandro, Diego Romero, Dolores FernÁndez-OrtuÑo, Francisco López-Ruiz, Antonio De Vicente, and Juan A. TorÉs. 2009. "The Powdery Mildew Fungus Podosphaera Fusca (Synonym Podosphaera Xanthii), a Constant Threat to Cucurbits: Pathogen Profile." *Molecular Plant Pathology*. https://doi.org/10.1111/j.1364-3703.2008.00527.x.

Pitrat, Michel. 2016. "Disease Resistance in Melon and Its Modification by Molecular Breeding Techniques." *Biotechnology in Agriculture and Forestry* 70. Springer International Publishing: 175–97. https://doi.org/10.1007/978-3-662-48535-4_11/TABLES/3.

Rahman, A., J. R. Standish, K. N. D'Arcangelo, and L. M. Quesada-Ocampo. 2021. "Clade-Specific Biosurveillance of Pseudoperonospora Cubensis Using Spore Traps for Precision Disease Management of Cucurbit Downy Mildew." *Phytopathology* 111 (2). American Phytopathological Society: 312–20. https://doi.org/10.1094/PHYTO-06-20-0231-R/ASSET/IMAGES/LARGE/PHYTO-06-20-0231-RF6-1613748563775.JPEG.

Reuveni, M., Eyal, H. and Cohen, Y. 1980. "Development of Resistance to Metalaxyl in Pseudoperonospora cubensis." *Plant Disease* 64: 1108–09.

Risser, G., Banihashemi, Z., Davis, D. 1976. "A Proposed Nomenclature of Fusarium Oxysporum f. sp. Melonis Races and Resistance Genes in Cucumis melo." *Phytopathology* 66: 1105–06.

Runge, Fabian, Young Joon Choi, and Marco Thines. 2011. "Phylogenetic Investigations in the Genus Pseudoperonospora Reveal Overlooked Species and Cryptic Diversity in the P. Cubensis Species Cluster." *European Journal of Plant Pathology* 129 (2). Kluwer Academic Publishers: 135–46. https://doi.org/10.1007/S10658-010-9714-X/FIGURES/1.

Savory, Elizabeth A., Leah L. Granke, Lina M. Quesada-Ocampo, Marina Varbanova, Mary K. Hausbeck, and Brad Day. 2011. "The Cucurbit Downy Mildew Pathogen Pseudoperonospora Cubensis." *Molecular Plant Pathology* 12 (3): 217–26. https://doi.org/10.1111/J.1364-3703.2010.00670.X.

SCGrower. (2022). "Resistance to Tebuconazole is Widespread in the Gummy Stem Blight Fungus in South Carolina." Accessed January 15, 2025. https://scgrower.com/2022/06/14/resistance-to-tebuconazole-is-widespread-in-the-gummy-stem-blight-fungus-in-south-carolina/.

Schroeder, W. T., Provvidenti, R. 1969. "Resistance to Benomyl in Powdery Mildew of Cucurbits." *Pl. Dis. Reptr*. 53: 271–75.

Sitterly, W. 1969. "Effect of Crop Rotation on Cucumber Gummy Stem Blight." *Plant Disease Report* 53: 417–19.

Sousaraei, Naeimeh, Hossein Ramshini, Mahmoud Lotfi, and Abbas Sharzei. 2018. "Marker Assisted Backcrossing for Introgression of Fusarium Wilt Resistance Gene into Melon." *Euphytica* 214 (1). Springer Netherlands. https://doi.org/10.1007/s10681-017-2080-4.

Stevenson, Katherine L., David B. Langston, Kenneth W. Seebold, and K. L. Langston. 2004. "Resistance to Azoxystrobin in the Gummy Stem Blight Pathogen Documented in Georgia." *Plant Health Progress* 5 (1). Scientific Societies. https://doi.org/10.1094/PHP-2004-1207-01-RS.

Stewart, J., Turner, A., Brewer, M. 2015. "Evolutionary History and Variation in Host Range of Three Stagonosporopsis Species Causing Gummy Stem Blight of Cucurbits." *Fungal Biology* 119 (5): 370–82. https://doi.org/10.1016/j.funbio.2014.12.008

Tezuka, T., K. Waki, M. Kuzuya, T. Ishikawa, Y. Takatsu, and M. Miyagi. 2011. "Development of New DNA Markers Linked to the Fusarium Wilt Resistance Locus Fom-1 in Melon." *Plant Breeding* 130 (2). John Wiley & Sons, Ltd: 261–67. https://doi.org/10.1111/J.1439-0523.2010.01800.X.

Thomas, C. 1986. "Downy and Powdery Mildew Resistant Muskmelon Breeding Line MR-1." *HortScience* 21 (2): 329–330. https://doi.org/10.21273/HORTSCI.21.2.329

Thomas, C. 1999. "Additional Evaluations of *Cucumis melo* L. Germplasm for Resistance to Downy Mildew." HortScience 920–921.

Thomas, C., Inaba, T. and Cohen, Y. 1987. "Physiological Specialization in Pseudoperonospora cubensis". *Phytopathology* 77: 1621–24.

Thomas, C., McCreight, J., Jourdain, E. 1990. "Inheritance of Resistance to Alternaria-Cucumerina in Cucumis-Melo Line MR-1." *Plant Disease* 74 (11): 868–70.

Thomas, C. and Jourdain, E. 1992. "Host Effect on Selection of Virulence Factors Affecting Sporulation by Pseudoperonospora cubensis." *Plant Disease* 76: 905–07.

Toporek, Sean M., Sandra E. Branham, Melanie L. Katawczik, Anthony P. Keinath, and W. Patrick Wechter. 2021. "QTL Mapping of Resistance to Pseudoperonospora Cubensis Clade 1, Mating Type A2, in Cucumis Melo." *Theoretical and Applied Genetics* 134: 2577–86. https://doi.org/10.1007/s00122-021-03843-w.

Toporek, Sean M., Sandra E. Branham, Anthony P. Keinath, and W. Patrick Wechter. 2023. "QTL Mapping of Resistance to Pseudoperonospora Cubensis Clade 2, Mating Type A1, in Cucumis Melo and Dual-Clade Marker Development." *Theoretical and Applied Genetics* 136: 91. https://doi.org/10.1007/s00122-023-04333-x.

Urban, J., and A. Lebeda. 2006. "Fungicide Resistance in Cucurbit Downy Mildew – Methodological, Biological and Population Aspects." *Annals of Applied Biology* 149 (1). John Wiley & Sons, Ltd: 63–75. https://doi.org/10.1111/J.1744-7348.2006.00070.X.

USDA, ERS. 2023. https://www.ers.usda.gov/sites/default/files/_laserfiche/outlooks/106240/FTS-376.pdf?v=25846. (accessed on 10 January 2025).

Vaughn, Justin N., Sandra E. Branham, Brian Abernathy, Amanda M. Hulse-Kemp, Adam R. Rivers, Amnon Levi, and William P. Wechter. 2022. "Graph-Based Pangenomics Maximizes Genotyping Density and Reveals Structural Impacts on Fungal Resistance in Melon." *Nature Communications* 13 (1). Nature Publishing Group: 1–14. https://doi.org/10.1038/s41467-022-35621-7.

Wallace, E. C., K. N. D'Arcangelo, and L. M. Quesada-Ocampo. 2020. "Population Analyses Reveal Two Host-Adapted Clades of Pseudoperonospora Cubensis, the Causal Agent of Cucurbit Downy Mildew, on Commercial and Wild Cucurbits." *Phytopathology* 110 (9). American Phytopathological Society: 1578–87. https://doi.org/10.1094/PHYTO-01-20-0009-R/ASSET/IMAGES/LARGE/PHYTO-01-20-0009-RT3-1597419473922.JPEG.

Wang, Shiwei, Jinghua Yang, and Mingfang Zhang. 2011. "Developments of Functional Markers for Fom-2-Mediated Fusarium Wilt Resistance Based on Single Nucleotide Polymorphism in Melon (*Cucumis melo* L.)." *Molecular Breeding* 27 (3). Kluwer Academic Publishers: 385–93. https://doi.org/10.1007/S11032-010-9439-8/TABLES/5.

Wang, Yi-Hong, Claude E. Thomas, and Ralph A. Dean. 2000. "Genetic Mapping of a Fusarium Wilt Resistance Gene (Fom-2) in Melon (*Cucumis melo* L.)." *Molecular Breeding* 6.

Wechter, W., Dean, R., Thomas, C. 1998. "Development of Sequence-Specific Primers that Amplify a 1.5-kb DNA Marker for Race 1 Fusarium Wilt Resistance in *Cucumis melo* L." *HortScience* 33: 291–92.

Wechter W, Whitehead M, Thomas C, Dean R. 1995. "Identifcation of a Randomly Amplifed Polymorphic DNA Marker Linked to the Fom2 Fusarium Wilt Resistance Gene in Melon MR-1." *Phytopathology* 85: 1245–49.

Yang, Jinghua, Guancong Deng, Jinmin Lian, Jenella Garraway, Yongchao Niu, Zhongyuan Hu, Jingquan Yu, and Mingfang Zhang. 2020. "The Chromosome-Scale Genome of Melon Dissects Genetic Architecture of Important Agronomic Traits." *IScience* 23 (8). Elsevier: 101422. https://doi.org/10.1016/J.ISCI.2020.101422.

Yao, Yanlai, Zhiyong Xue, Chunlai Hong, Feng Xiang Zhu, Xiaoyang Chen, Weiping Wang, Zucong Cai, Nan Huang, and Xinqin Yang. 2016. "Efficiency of Different Solarization-Based Ecological Soil Treatments on the Control of Fusarium Wilt and Their Impacts on the Soil Microbial Community." *Applied Soil Ecology* 108 (December). Elsevier: 341–51. https://doi.org/10.1016/J.APSOIL.2016.09.015.

Yuste-Lisbona, Fernando J., Carmen Capel, Emilio Sarria, Rocío Torreblanca, María L. Gómez-Guillamón, Juan Capel, Rafael Lozano, and Ana I. López-Sesé. 2011. "Genetic Linkage Map of Melon (*Cucumis melo* L.) and Localization of a Major QTL for Powdery Mildew Resistance." *Molecular Breeding* 27 (2). Kluwer Academic Publishers: 181–92. https://doi.org/10.1007/S11032-010-9421-5/FIGURES/2.

Zhao, Juan, Shuwen Bao, Guoping Ma, and Xuehong Wu. 2016. "Characterization of Alternaria Species Associated with Muskmelon Foliar Diseases in Beijing Municipality of China." *Journal of General Plant Pathology* 82 (1). Springer-Verlag Tokyo: 29–32. https://doi.org/10.1007/S10327-015-0631-X/FIGURES/2.

Zhang T, Xu N, Amanullah S and Gao P. 2023. "Genome-wide Identification, Evolution, and Expression Analysis of MLO Gene Family in Melon (*Cucumis melo* L.)." *Front. Plant Sci.* 14: 1144317. https://doi.org/10.3389/fpls.2023.1144317

Zhou, X. G., and K. L. Everts. 2008. "First Report of Alternaria Alternata f. Sp. Cucurbitae Causing Alternaria Leaf Spot of Melon in the Mid-Atlantic Region of the United States." *The American Phytopathological Society* 92 (4): 652. https://doi.org/10.1094/PDIS-92-4-0652B.

Zink, F. 1992. "Genetics of Resistance to *Fusarium oxysporum* f. sp. *melonis* races 0 and 2 in Muskmelon Cultivars Honey Dew, Iroquois, and Delicious 51." *Plant Disease* 76: 162–66.

Zink, F., Thomas, C. 1990. "Genetics of Resistance to *Fusarium oxysporum* f. sp. *melonis* races 0, 1, and 2 in melon line MR-1." *Phytopathology* 80: 1230–32.

Zuniga, T., Jantz, J., Zitter, T., Jahn, M. 1999. "Monogenic Dominant Resistance to Gummy Stem Blight in Two Melon (Cucumis melo) Accessions." *Plant Disease* 83 (12): 1105–07.

Zuniga, T. L., T. A. Zitter, T. R. Gordon, D. T. Schroeder, and D. Okamoto. 1997. "Characterization of Pathogenic Races of Fusarium Oxysporum f. Sp. Melonis Causing Fusarium Wilt of Melon in New York." *Plant Disease* 81.

Zhang, Y., Kyle, M., Anagnostou, K., Zitter, T. 1997. "Screening Melon (Cucumis melo) for Resistance to Gummy Stem Blight in the Greenhouse and Field." HortScience 32 (1): 117–21.

7 Breeding for High Yield and Fruit Quality in Melon

Rakesh Kumar

7.1 INTRODUCTION

Melon (*Cucumis melo* L.) is a widely cultivated member of the Cucurbitaceae family, which also encompasses watermelons, cucumbers, gourds, pumpkins, and squash. It is diploid ($2n = 2x = 24$) and has a genome size of 454 Mb (Garcia-Mas et al. 2012). Melon is a highly diverse crop in terms of plant or fruit-related traits such as typology, size, shape, sex expression, and flesh color. Prominent types of melon include cantaloupe, honeydew, Piel de Sapo, yellow canary, Hami, Charentais, Italo-American, and Ananas.

Melons are cultivated on nearly every continent, with their genetic diversity contributing to biodiversity and sustainable agricultural practices. In 2023, melon production covered 1.09 million hectares, yielding 29.5 million tonnes of fruit (FAOSTAT 2025). Cultivation in Asia accounted for 75% of the total acreage, followed by the Americas (13%), Europe (6.5%), and Africa (4.4%). Within Asia, China is the largest producer (14 million tonnes), followed by India (1.5 million tonnes) and Türkiye (1.4 million tonnes). Other prominent melon-producing nations include Kazakhstan, Afghanistan, Guatemala, Iran, Spain, Brazil, and Italy (FAOSTAT 2025).

While melons are grown in many diverse agro-climatic zones, they are highly sensitive to temperature, day length, humidity, light intensity, and soil conditions. Hence, breeding efforts should aim to develop cultivars that are adapted to local growing conditions. Melons' high genetic diversity and differences in local consumer preferences make them an ideal candidate for breeding programs. Breeding efforts are primarily focused on traits favored by growers, including yield, fruit quality, post-harvest shelf life, and resistance to both abiotic and biotic stresses. Fruit quality is determined mainly by flesh firmness, aroma, flavor, shelf life, and total soluble solids (TSSs). Key abiotic stressors affecting melons include heat stress, cold stress, fruit cracking, and fruit rotting. In addition, melons are challenged by biotic stresses, such as powdery mildew, downy mildew, Fusarium wilt, anthracnose, viral diseases, aphids, whiteflies, thrips, and beetles. Breeders have made tremendous progress in improving tolerance to abiotic and biotic stresses, as many related traits are simply inherited and controlled by one or a few genes (qualitative).

While agricultural productivity has improved significantly over the years, population growth and climate change continue to threaten global food security (Wheeler

DOI: 10.1201/9781003715757-7

and von Braun 2013; Gerland et al. 2014). Therefore, genetic improvements of crop yields are critical for ensuring sustainable food production. Yield and yield-related traits are difficult to improve through breeding due to their complex inheritance (quantitative). Similar considerations apply to fruit quality traits, such as flavor and aroma. Because quantitative traits are strongly influenced by the environment and exhibit low heritability, breeders must consider local conditions when developing high-yielding varieties. Yield is primarily controlled by the non-additive components of genetic variance, which are not directly heritable. This deviation from additive variance leads to heterosis, also known as hybrid vigor, which is a significant contributor to crop yield improvement (East 1908). Heterosis is known to significantly impact crop yields, resulting in estimated increases of 15%–30% (Duvick 2001). For instance, heterosis has driven continuous yield gains in maize since the introduction of the first commercial hybrid in the 1930s (Duvick 2001; Troyer 2006).

7.2 HETEROSIS

Melon breeders strive to develop cultivars with superior yields, fruit quality, and stress resistance. Given that most melon cultivars are hybrids, heterosis breeding remains the most widely used strategy for improving fruit yield and quality (Banga and Banga 2000). While the underlying mechanisms are poorly understood, heterosis, which refers to the superiority of hybrids over their parents, has been exploited in crop breeding for decades. Heterosis is highly complex and can vary significantly between parental lines, traits of interest, and environmental conditions (Napolitano et al. 2020). However, advancements in genomics, proteomics, transcriptomics, and metabolomics are enabling researchers to explore the scientific basis for the phenomenon.

Melon fruit yield varies substantially among cultivars (between 6,101 and 25,173 kg per hectare) and is influenced by genetics, environmental factors, and cultivation practices. A study of 85 diverse melon varieties reported that the average yield per plant ranged from 2.47 to 6.76 kg (Ari et al. 2022).

Several studies have documented heterosis for fruit yield in melon (Kesh and Kaushik 2021). For instance, Munger (1942) reported a 30% yield increase in hybrids when compared with their parents. Heterosis is commonly measured through diallel analysis, which estimates the general combining ability (GCA) of parents and specific combining ability (SCA) of hybrids (Barros et al. 2011). GCA reflects additive and additive × additive genetic effects and is used to identify parents with a high frequency of favorable alleles (Shattuck et al. 1993). Conversely, SCA represents dominance, additive × dominance, and dominance × dominance effects and is used to identify promising hybrid combinations (Shattuck et al. 1993; Valério et al. 2009).

GCA and SCA have been extensively studied in relation to yield in melon. Barros et al. (2011) observed significant GCA and SCA effects on yield, flesh firmness, and TSSs. Singh and Randhawa (1990) and Zalapa et al. (2006) documented that additive effects influence fruit number. Lippert and Legg (1972) and Abadia et al. (1985) further found that yield is impacted by both GCA and SCA, with the predominance of non-additive variance associated with the quadratic component of SCA. Dafna et al. (2021) reported that hybrid fruit yield was, on average, 73% higher than that of their parental lines. In Piel de Sapo melon, José et al. (2005) observed no heterosis for soluble solid concentration, negative to positive heterosis for fruit weight and diameter, and a general positive heterosis for ovary shape, fruit length, and fruit shape. Napolitano et al. (2020) identified additive gene effects as the primary components contributing to yield per plant, TSSs, and earliness. High levels of additive gene effects are associated with high narrow-sense heritability, suggesting that optimal genetic gains can be achieved through selective breeding. Many studies have confirmed the additive gene effects on yield per plant (Pouyesh et al. 2017).

In addition to fruit number, GCA and SCA have been found to modulate melon fruit weight. For instance, Shashikumar and Pitchaimuthu (2016) and Pouyesh et al. (2017) observed both additive and non-additive effects, finding that the former were more prominent. However, while examining a cross between the strain of growth habits (Fractal) 'USDA 846–1' and the cantaloupe cultivar 'Top Mark', Zalapa et al. (2006) observed that the inheritance of average fruit weight is governed by dominance and epistasis effects. Furthermore, Monforte et al. (2004) and Paris et al. (2008) used molecular markers to verify the polygenic control of average fruit weight.

Additive and non-additive effects also govern critical agronomic traits in melon that are unrelated to yield. Several significant studies have demonstrated the influence of both GCA and SCA on flesh firmness, a vital characteristic for long-distance shipping (Barros et al. 2011; Shashikumar and Pitchaimuthu 2016; Dantas et al. 2023). For TSSs, a key component of fruit flavor, Dantas et al. (2023) identified additive and non-additive effects, whereas Monforte et al. (2005) implicated a single significant additive effect. Studies on yield per plant revealed that GCA sources were consistently substantial, whereas SCA was never significant, indicating that this trait is controlled solely by additive gene action (Napolitano et al. 2020). Esmaeili et al. (2022) reported substantial GCA for fruit number per plant, fruit yield per plant, fruit weight, flesh firmness, and TSSs. However, SCA effects exceeded GCA for fruit yield per plant, indicating the importance of dominance and non-additive gene actions for this trait. The high narrow-sense heritability for yield per plant (0.86), fruit number (0.70), and flesh firmness (0.72) indicates that these traits are minimally influenced by environmental pressures and can be effectively targeted through selection. Conversely, TSSs exhibit a low, narrow-sense heritability (0.38), suggesting a higher environmental impact. In a study comprising 49 genotypes, Badami et al. (2020) found that SCA significantly influenced fruit weight, TSSs, fruit length, and fruit diameter. In contrast, GCA had no substantial effect on these traits. This result suggests that dominant alleles are responsible for controlling these traits. A diallel study of seven genotypes by Shajari et al. (2021) found a positive correlation between fruit weight and fruit yield, along with significant GCA and SCA estimates for total yield, fruit number, TSSs, flesh firmness, and flesh thickness. Similar results were reported by Feyzian et al. (2009) and Pouyesh et al. (2017). In contrast, Barros et al. (2011) observed no notable SCA for fruit weight, but documented significant GCA effects and high GCA/SCA ratios for TSSs, fruit shape, flesh firmness, and thickness, indicating the high potential for trait selection. In their 2021 study of diallel-developed melon hybrids, Dafna et al. reported both rootstock-mediated and best-parent yield heterosis, with the latter averaging over 25%. GGE (genotype + genotype × environment) biplots were first developed by Yan (1999) and Yan et al. (2000) to investigate genotypic and environmental interactions for yield and yield-related traits. Later, Yan and Hunt (2002) proposed using this method to analyze diallel data for cereal crops. Dehghani et al. (2012) were the first to apply GGE biplots to analyze yield-related traits in melon populations.

7.2.1 Heterosis and Genetic Distance

First proposed by Melchinger (1999), the genetic distances derived from molecular markers offered a promising tool for grouping germplasms with similar genomes and for identifying heterotic patterns. The majority of studies conducted in maize revealed a clear, positive correlation between the genetic divergence of parental lines and potential heterosis. However, Tomkowiak et al. (2019) suggested that the magnitude of the phenomenon may be lower than expected. Thus, while genetic divergence is a necessary condition, it is not a guarantee of heterotic performance (Kaeppler 2012). In melon, Luan et al. (2010) reported that, despite differences in performance between parents and among F_1 hybrids, a significant correlation between genetic distance and heterotic effect was only evident for branch number.

Kalb and Davis (1984) hypothesized that the geographic and genetic distance among American melon cultivars was positively correlated with high performance. In an attempt to correlate genetic distance with heterosis, Napolitano (2015) suggested that molecular markers could be used for parent selection when pedigree data are not available. However, several studies have been unable to find a correlation between genetic distance and heterosis, even when combining traits in a multivariate dimension (Mahalanobis and Euclidean distances; Burstin and Charcosset 1997; Heckenberger et al. 2005; Teklewold and Becker 2006). This result may be due to the assessed markers indicating neutral genomic regions rather than the traits of interest. While this can be a viable approach for pre-breeding studies, particularly when no markers are associated with the desired characteristic or when the influence of SCA is negligible (i.e., TSSs), these studies highlight the challenges of predicting hybrid performance and heterosis, even with well-characterized parental lines. The lack of correlation between genetic distance and heterosis limits the utility of genomic tools for predicting hybrid yield (Kaushik et al. 2018; Napolitano et al. 2020).

Next-generation sequencing technologies and the growing availability of whole-genome assemblies are providing new tools for studying heterosis. Ongoing efforts aim to utilize new technologies to explore the molecular basis of heterosis in model and crop plants (Huang et al. 2016; Li et al. 2016; Seymour et al. 2016; Yang et al. 2017). For example, to investigate rootstock heterosis for yield improvement, Dafna et al. (2021) conducted whole-genome resequencing of 20 founder lines and found that parental genetic distance is not a predictor of heterosis.

7.3 GENOMICS OF MELON FRUIT CHARACTERISTICS

7.3.1 Yield

Fruit yield is a key concern for melon growers, as they are paid by the ton. The global production of melons averages 29.5 million tons annually (FAOSTAT 2025).

Melon yields vary among cultivars and environments, and are directly related to fruit weight, length, width, and shape (Rad et al. 2010; Ari et al. 2022). Each melon can weigh between a few grams and 15 kg (Goldman 2002). Secondary traits such as plant height, number of branches, flowers, and fruits per plant are indirectly associated with fruit yield. Additionally, researchers have identified several quantitative trait loci (QTLs) and candidate genes related to fruit yield (Zhang et al. 2020). In their study, Zhang et al. (2020) identified 28 QTLs associated with fruit traits, including fruit weight, length, width, and TSSs, concluding that the genetic components for fruit quality are complex and additive. They constructed a linkage map using 195 CAPS markers in an F_2 population derived from the thick-skinned line 'Elizabeth' (M4–5) and a thin-skinned line, M1–15, revealing that most QTLs associated with fruit weight and length were clustered on chromosomes 6 and 9 (Zhang et al. 2020). Similarly, Amanullah et al. (2021) also identified QTLs for melon yield–related traits (fruit weight, length, and width) on chromosome 9 using QTL mapping with SNP-based CAPS markers and whole genome resequencing. Amanullah et al. (2022) later employed SNP-derived CAPS markers to map QTLs FWt2.1, FWt4.1, and FWt9.1 for fruit weight, as well as one major QTL for fruit length (FL12.1).

An extensive study by Lian et al. (2021) utilized whole-genome sequencing to construct two high-density genetic maps from two populations of *Cucumis melo* ssp. *agrestis* var *chinensis* x Cucumis melo ssp. *agrestis* var *agrestis* (WAP) and *Cucumis melo* ssp. *melo* var *chandalak* x *Cucumis melo* ssp. *agrestis* var *conomon* (MAP). A total of 1,871,671 and 1,976,589 high-quality SNPs were identified in the WAP and MAP populations, respectively. The results revealed two key loci influencing melon fruit size: one on chromosome 11 in WAP and one on chromosome 5 MAP. In the WAP population, two QTLs linked to fruit diameter and weight were located in the same 130.8 Kb region on chromosome 11. This overlap highlighted *MELO3C025758*, which encodes an auxin response factor (ARF), as a possible candidate gene involved in determining melon fruit size. In the MAP population, a region on chromosome 5 was found to impact fruit weight and length. This region was found to contain the gene *MELO3C004493*, which encodes a YABBY transcription factor (a type known to influence tomato fruit size). As a result, both *MELO3C025758* and *MELO3C004493* are proposed candidate genes for modulating melon fruit size.

Previous studies have found a positive correlation between fruit yield and size (weight; Zalapa et al. 2007). Ramamurthy and Waters (2015) reported that a small segment on linkage group 8 may be responsible for yield and yield-related traits. This finding is further supported by several QTLs in linkage group 8 associated with fruit weight and fruit number, which were identified in a RIL population created from USDA 846–1 × 'Top Mark' (Zalapa et al. 2007). Regarding average fruit weight, Dantas et al. (2023) estimated approximately 25 and 10 loci in OL × A-16 and OL × PV crosses, respectively. Meanwhile, Zalapa et al. (2006) reported negative estimates for the number of loci of the average fruit weight. However, average fruit weight has been indicated as a polygenic trait (Monforte et al. 2004; Eduardo et al. 2007; Wang et al. 2016; Amanullah et al. 2018). Monforte et al. (2004) identified six QTLs from the F_2 and double-haploid generation of 'Santa Claus' × PI accession 161375 crossing. Eduardo et al. (2007) further studied isogenic strains derived from this crossing, uncovering 11 QTLs controlling the average fruit weight. Later, a cross between

Védrantais and Piel de Sapo revealed *FWQU5.1* and *FWQU8.1*, two QTLs that control fruit weight (Pereira et al. 2018). *FWQU5.1* contains genes responsible for fruit size regulators and increased fruit weight when the *'Védrantais'* allele was present. Finally, Dafna et al. (2021) reported that two central root-mediated QTLs, q.RMY3.1 and q.RMY6.1, increase yield and display heterotic inheritance.

7.3.2 Total Soluble Solids (TSS)

TSS content is one of the most important traits to consider when developing improved melon varieties. Consumers strongly consider the sweetness of melon, and many retailers have established a minimum sugar threshold for produce acceptability. Sucrose, fructose, and glucose are the three primary sugars found in melons. Sucrose is the most dominant sugar, and its rapid accumulation during the final stages of maturity determines the sweetness of a melon.

Understanding the genetics of sugar accumulation is a top priority for melon breeders. Over the years, several QTLs associated with sugar content have been reported in diverse genotypes (Monforte et al. 2004; Harel-Beja et al. 2010). Burger et al. (2002) reported that a single major gene, *suc*, has a significant influence on sucrose accumulation. By merging published QTLs from 18 mapping trials, Diaz et al. (2011) constructed a consensus linkage map that included more than 10 QTLs related to Brix values, sucrose, and sugar content. An additional mapping study by Harel-Beja et al. (2010) reported six significant QTLs for sugar content on chromosome 2. Additionally, the study revealed one QTL for glucose content on chromosome 4 and determined that the low-sugar parent alleles contributed to a higher glucose content, resulting in transgressive segregation.

Final sugar accumulation in melon is a function of sucrose synthesis. The metabolic pathway of sucrose accumulation is regulated by 42 genes and involves several key enzymes, including invertases, sucrose synthase, and sucrose phosphate synthase (Xu et al. 2022). A study by Dantas et al. (2023) reported between six and eight loci associated with TSSs. Monforte et al. (2004) identified five QTLs from the crossing between a Santa Claus strain and a Korean accession (PI 161375). Isogenic strains derived from this crossing were employed by Eduardo et al. (2007) to identify a total of 15 QTLs. Finally, Melo et al. (2011) recorded 28 loci involved in sugar accumulation in a cross between *inodorus* (honeydew) and *acidulous* melon strains.

Recently, a cross between 'Piel de Sapo' and 'Sogwhan Charmi' uncovered three important clusters related to sugar content on chromosomes 4, 5, and 7 (Argyris et al. 2017). Among them, a major QTL, *SUCQSC5.1*, reduced TSS and sucrose content, for which *MELO3C014519* was considered the candidate gene. In addition, Schemberger et al. (2020) detected differentially expressed genes responsible for sucrose content in melon, including *CmINH3*, *CmTPP1*, *CmTPS5*, *CmTPS7*, and *CmTPS9*. Because sugars are made in mesophyll cells and transported to other parts of the plant, accumulation in melon fruit depends on both sugar synthesis and transport (Xu et al. 2022). Cheng et al. (2018) isolated three tonoplast sugar transporters, *CmTST1*, *CmTST2*, and *CmTST3*. Of these, *CmTST2* plays an essential role in sugar accumulation in melon. A meta-genetic study conducted by Tzuri et al. (2025) analyzed melon sweetness using six populations and 15 published melon sweetness-related QTL studies.

They found that sucrose accumulation was responsible for TSS variation and exhibited additive effects. Furthermore, they identified several QTLs associated with TSS in regions of chromosome 12. In their cross between 'Védrantais' (*cantalupensis*) and 'Piel de Sapo' (*inodorus*), Pereira et al. (2018) detected six QTLs for TSS on chromosomes 8, 9, and 10. Among them, QTL SSCQU8.3 on chromosome 8 had the most significant effect.

Although many QTLs associated with sugar accumulation in melon have been detected, the diverse genetic backgrounds of the analyzed melon populations complicate the selection of candidate genes for validation. Moreover, sugar accumulation is a highly complex trait influenced by both genetic background and growth environment.

7.3.3 Firmness

Flesh firmness is directly responsible for both field performance and post-harvest shelf life. Melon is categorized as having a medium, extended, or long shelf life. Melons cultivated for long-distance shipping require firm flesh, which can maintain stability for over a month. Meanwhile, melons consumed locally only require medium flesh firmness.

Harel-Beja et al. (2010) identified two QTLs for flesh firmness in a RIL population derived from PI414723 × 'Dulce'. In their study, Dahmani-Mardas et al. (2010) constructed a mutant collection of 4,023 melon M_2 TILLING families and cloned genes involved in ethylene production. They then identified *L124F* and *G194D*, two missense mutations of the ethylene biosynthetic enzyme ACC oxidase 1. *G194D* mutant fruit displayed delayed ripening and an enhanced shelf life. In a genome-wide association study of 125 accessions and an F_2-segregating population between MR-1 and Hale's Best Jumbo, Nimmakayala et al. (2016) revealed 11 SNPs across chromosomes 6, 8, 11, and 12 that were associated with fruit firmness. Dantas et al. (2023) reported six to eight loci in OL × A-16 and OL × PV populations. Further gene annotation revealed that several SNPs on these loci are associated with β-D-xylosidase, glyoxysomal malate synthase, chloroplastic anthranilate phosphoribosyl transferase, and histidine kinase. These genes have been previously characterized regarding their roles in fruit ripening and softening in other crops (Zhang et al., 2016).

Near-isogenic lines derived from the non-climacteric parental melon lines PI 161375 and 'Piel de Sapo' were used by Moreno et al. (2008) to map QTLs colocalizing with candidate genes involved in ethylene regulation, biosynthesis, and cell wall degradation. These included *CmACS5* on linkage group III, *CmEXP3* and *CmXTH5* on linkage group ×, and *CmEIL1*, *CmEIL3,* and *CmEIL4* on linkage groups II and III.

7.3.4 Fruit Shape

Fruit shape significantly impacts both commercial appeal and pre- and post-harvest handling, making it a key consideration in melon breeding programs. While most commercial cantaloupes are round, varieties like Piel de Sapo, Yellow Canary, and Hami are elongated. A single gene controls the elongated fruit shape, which is dominant over the round shape. This notion is supported by Bains and Kang (1963), who identified the recessive *sp* gene as responsible for round fruit shape. Normally, monecious (*A-G-*) cultivars are elongated, and andromonoecious (*aaG-*) ones are round, which may be due to a fruit shape QTL that colocalizes with the andromonoecious locus on chromosome 2. The gene *CmACS7*, located on chromosome 2, is involved in the ethylene biosynthesis pathway and regulates both monecious sex expression and elongated fruit shape (Boualem et al. 2022).

The shape of the melon fruit is also influenced by the number of carpels, with pentamerous fruit tending to be rounder than trimerous fruit. A QTL associated with rounder fruit is known to colocalize with the *p* gene (pentamerous) on chromosome 12. In andromonoecious fruit, shape is altered by the *CmCLV3* gene, which determines the number of carpels (Boualem et al. 2008; Liu et al. 2020). Additionally, a QTL on chromosome 8 harbors a gene encoding an OVATE family protein (OFP), which has been identified as a candidate for influencing fruit shape. The elevated expression of this gene, caused by a retrotransposon insertion in the promoter, was associated with fruit development in a flat-round HP22 line (Ari et al. 2022).

On chromosome 6, QTLs *FLQU6.1* and *FSQU6.2* resulted in a decrease in fruit length caused by the Védrantais allele, thereby decreasing the shape index (Pereira et al. 2018). *FLQU6.1* and *FSQU6.2* also colocalize with a recently published QTL correlated with the same traits, which was mapped in an F_2 population between PS and the Indian wild accession 'Trigonus' (Díaz et al. 2017). Pereira et al. (2018) documented 12 candidate genes within the QTL intervals of *FSQU2.1*, *FDQU2.1*, *FSQU6.1*, *FSQU6.2*, *FWQU8.1*, and *FSQU11.1*. Four of these genes (*MELO3C015418*, *MELO3C025343*, *MELO3C013751*, and *MELO3C022253*) displayed expression patterns consistent with roles in the ovaries and young fruit.

7.3.5 Flavor and Aromatic Compounds

Melons are prized for their refreshing taste and aroma during the hot summer months. Flavor and aroma vary greatly among melon typologies, with distinct profiles among honeydews, Charentais, cantaloupes, and Galias. The primary volatile aromatic compounds include esters, aldehydes, alcohols, ketones, and terpenes (Shahwar et al. 2023). The presence and concentration of these compounds differ between melons. For example, ethyl-2-methyl butyrate is prominent in cantaloupes (Allwood et al. 2014), while (E,Z)-nonadienal is characteristic of honeydews (Aubert and Bourger 2004). Lipoxygenases and hydroperoxide lyase are involved in the formation of aromatics in melon (Tang et al. 2015). Zhang et al. (2014) reported 18 lipoxygenase genes in oriental melon, suggesting that specific LOX genes are essential for volatile compounds. While investigating a cross between PI414723 and 'Dulce' cantaloupe, Galpaz et al. (2018) identified a major QTL that contained the gene *CmThAT1*. This gene encodes S-methyl-thioacetate, which is a key component of melon fruit aroma.

In another study, a cross between 'Piel de Sapo' (*inodorus* type) and 'Védrantais' (*cantalupensis*) revealed numerous QTLs, including one on chromosome 8 that is linked to ethylene-related biosynthesis (Mayobre et al. 2021). Meanwhile, Perpiñá et al. (2021) identified 57 volatile compounds in 25 introgression lines derived from the introgression of 'Ginsen Makuwa' into the Charentais background of the cultivar

'Védrantais'. The joint analysis of these compounds using genotype by sequencing helped identify candidate genes. In particular, candidate gene *MELO3C024886* (a 4-coumarate: CoA ligase) is likely involved in the eugenol biosynthesis pathway. Other candidate genes (*CmLOX18*, *CmAAT*, *CmBAMT*, and *CmCRTISO1*) have been previously implicated in the production of volatile compounds from fatty acids, amino acids, and carotenoids.

REFERENCES

Abadia, J., M. L. Cuarteiro, and F. Nuez. 1985. *Herencia de Caracteres Cuantitativos en Melón*. Vol. 28. Instituto Nacional de Investigación y Tecnología Agraria y Alimentaria, 83–91.

Allwood, J. William, William Cheung, Yun Xu, Roland Mumm, Ric C. H. De Vos, Catherine Deborde, Benoit Biais, Mickael Maucourt, Yosef Berger, Arthur A. Schaffer, Dominique Rolin, Annick Moing, Robert D. Hall, and Royston Goodacre. 2014. "Metabolomics in Melon: A New Opportunity for Aroma Analysis." *Phytochemistry* 99: 61–72. https://doi.org/10.1016/j.phytochem.2013.12.010.

Amanullah, Sikandar, Peng Gao, Benjamin Agyei Osae, Arvind Saroj, Tiantian Yang, Shi Liu, Yiqun Weng, and Feishi Luan. 2021. "Genetic Linkage Mapping and QTLs Identification for Morphology and Fruit Quality Related Traits of Melon by SNP Based CAPS Markers." *Scientia Horticulturae* 278: 109849. https://doi.org/10.1016/j.scienta.2020.109849.

Amanullah, Sikandar, Shi Liu, Peng Gao, Zicheng Zhu, Qianglong Zhu, Chao Fan, and Feishi Luan. 2018. "QTL Mapping for Melon (*Cucumis melo* L.) Fruit Traits by Assembling and Utilization of Novel SNPs Based CAPS Markers." *Scientia Horticulturae* 236: 18–29. https://doi.org/10.1016/j.scienta.2018.02.041.

Amanullah, Sikandar, Benjamin Agyei Osae, Tiantian Yang, Farhat Abbas, Shi Liu, Hongyu Liu, Xuezheng Wang, Peng Gao, and Feishi Luan. 2022. "Mapping of Genetic Loci Controlling Fruit Linked Morphological Traits of Melon Using Developed CAPS Markers." *Molecular Biology Reports* 49 (6): 5459–72. https://doi.org/10.1007/s11033-022-07263-x.

Argyris, Jason M., Aurora Díaz, Valentino Ruggieri, Marta Fernández, Torben Jahrmann, Yves Gibon, Belén Picó, Ana M. Martín-Hernández, Antonio J. Monforte, and Jordi Garcia-Mas. 2017. "QTL Analyses in Multiple Populations Employed for the Fine Mapping and Identification of Candidate Genes at a Locus Affecting Sugar Accumulation in Melon (*Cucumis melo* L.)." *Frontiers in Plant Science* 8: 1679. https://doi.org/10.3389/fpls.2017.01679.

Ari, B. Ç., N. Kayak, K. A. L. Ünal, I. S. S. I. Neslihan, and Ö. Türkmen; Y. Dal, E. S. Kurtar, M. Seymen. 2022. "Determination of Yield and Quality Characteristics of Hybrid Melon Cultivar Candidates in Konya Ecology." *Mustafa Kemal University Journal of Agricultural Sciences* 27: 309–17.

Aubert, Christophe, and Nelly Bourger. 2004. "Investigation of Volatiles in Charentais Cantaloupe Melons (Cucumis melo Var. Cantalupensis). Characterization of Aroma Constituents in Some Cultivars." *Journal of Agricultural & Food Chemistry* 52 (14): 4522–8. https://doi.org/10.1021/jf049777s.

Badami, K., B. Daryono, and S. Khoiri. 2020. "Combining Ability and Heterotic Studies on Hybrid Melon (*Cucumis melo* l.) Populations for Fruit Yield and Quality Traits." *SABRAO Journal of Breeding & Genetics* 52: 402–17.

Bains, M. S. and U. S. Kang. 1963. "Inheritance of some flower and fruit characters in muskmelon." *Indian Journal of Genetics and Plant Breeding* 23: 101–6.

Banga, S. S., and S. K. Banga. 2000. *Hybrid Cultivar Development*. New Delhi: Narosa Publishing House.

Barros, Anne Katherine de Araújo, Glauber Henrique de Sousa Nunes, Manoel Abílio de Queiróz, Elaíne Welk Lopes Pereira, and Jose Hamilton da Costa Filho. 2011. "Diallel Analysis of Yield and Quality Traits of Melon Fruits." *Crop Breeding & Applied Biotechnology* 11 (4): 313–9. https://doi.org/10.1590/S1984-70332011000400004.

Boualem, Adnane, Serge Berthet, Ravi S. Devani, Celine Camps, Sebastien Fleurier, Halima Morin, Christelle Troadec, Nathalie Giovinazzo, Nebahat Sari, Catherine Dogimont, and Abdelhafid Bendahmane. 2022. "Ethylene Plays a Dual Role in Sex Determination and Fruit Shape in Cucurbits." *Current Biology: CB* 32 (11): 2390–401.e4. https://doi.org/10.1016/j.cub.2022.04.031.

Boualem, Adnane, Mohamed Fergany, Ronan Fernandez, Christelle Troadec, Antoine Martin, Halima Morin, Marie-Agnes Sari, Fabrice Collin, Jonathan M. Flowers, Michel Pitrat, Michael D. Purugganan, Catherine Dogimont, and Abdelhafid Bendahmane. 2008. "A Conserved Mutation in an Ethylene Biosynthesis Enzyme Leads to Andromonoecy in Melons." *Science* 321 (5890): 836–8. https://doi.org/10.1126/science.1159023.

Burger, Yosef, Uzi Saar, Nurit Katzir, Harry S. Paris, Yelena Yeselson, Ilan Levin, and Arthur A. Schaffer. 2002. "A Single Recessive Gene for Sucrose Accumulation in Cucumis melo Fruit." *Journal of the American Society for Horticultural Science* 127 (6): 938–43. https://doi.org/10.21273/JASHS.127.6.938.

Burstin, Judith, and Alain Charcosset. 1997. "Relationship between Phenotypic and Marker Distances: Theoretical and Experimental Investigations." *Heredity* 79 (5): 477–83. https://doi.org/10.1038/hdy.1997.187.

Cheng, Jintao, Suying Wen, Shuang Xiao, Baiyi Lu, Mingru Ma, and Zhilong Bie. 2018. "Overexpression of the Tonoplast Sugar Transporter CmTST2 in Melon Fruit Increases Sugar Accumulation." *Journal of Experimental Botany* 69 (3): 511–23. https://doi.org/10.1093/jxb/erx440.

Dafna, Asaf, Ilan Halperin, Elad Oren, Tal Isaacson, Galil Tzuri, Ayala Meir, Arthur A. Schaffer, Joseph Burger, Yaakov Tadmor, Edward S. Buckler, and Amit Gur. 2021 Sept. 30. "Underground Heterosis for Yield Improvement in Melon." *Journal of Experimental Botany* 72 (18): 6205–18. https://doi.org/10.1093/jxb/erab219, https://pubmed.ncbi.nlm.nih.gov/33993257.

Dahmani-Mardas, Fatima, Christelle Troadec, Adnane Boualem, Sylvie Lévêque, Abdullah A. Alsadon, Abdullah A. Aldoss, Catherine Dogimont, and Abdelhafid Bendahmane. 2010. "Engineering Melon Plants with Improved Fruit Shelf Life Using the TILLING Approach." *PLoS One* 5 (12): e15776. https://doi.org/10.1371/journal.pone.0015776.

Dantas, Ana C. A., Anânkia O. Ricarte, Jose M. da Costa, Rafaela P. Antônio, Francisco L. S. Tomaz, and Glauber H. S. Nunes. 2023. "Genetic Control of Quality of Melon Traits." *Ciência Rural* 53 (7): e20220089. https://doi.org/10.1590/0103-8478cr20220089.

Dehghani, Hamid, Ehsan Feyzian, Mokhtar Jalali, Abdolmajid Rezai, and Fenny Dane. 2012. "Use of GGE Biplot Methodology for Genetic Analysis of Yield and Related Traits in Melon (*Cucumis melo* L.)." *Canadian Journal of Plant Science* 92 (1): 77–85. https://doi.org/10.4141/cjps2010-046.

Diaz, Aurora, Mohamed Fergany, Gelsomina Formisano, Peio Ziarsolo, José Blanca, Zhanjun Fei, Jack E. Staub, Juan E. Zalapa, Hugo E. Cuevas, Gayle Dace, Marc Oliver, Nathalie Boissot, Catherine Dogimont, Michel Pitrat, René Hofstede, Paul van Koert, Rotem Harel-Beja, Galil Tzuri, Vitaly Portnoy, Shahar Cohen, Arthur Schaffer, Nurit Katzir, Yong Xu, Haiying Zhang, Nobuko Fukino, Satoru Matsumoto, Jordi Garcia-Mas, and Antonio J. Monforte. 2011. "A Consensus Linkage Map for Molecular Markers and Quantitative Trait Loci Associated with Economically Important Traits in Melon (*Cucumis melo* L.)." *BMC Plant Biology* 11: 111. https://doi.org/10.1186/1471-2229-11-111.

Díaz, Aurora, Ana Montserrat Martín-Hernández, Ramón Dolcet-Sanjuan, Ana Garcés-Claver, José María Álvarez, Jordi Garcia-Mas, Belén Picó, and Antonio José Monforte. 2017. "Quantitative Trait Loci Analysis of Melon (*Cucumis melo* L.) Domestication-Related

Traits." *Theoretical & Applied Genetics* 130 (9): 1837–56. https://doi.org/10.1007/s00122-017-2928-y.

Duvick, D. N. 2001. "Biotechnology in the 1930s: The Development of Hybrid Maize." *Nature Reviews. Genetics* 2 (1): 69–74. https://doi.org/10.1038/35047587.

East, E. M. 1908. "Inbreeding in Corn." In *Reports of the Connecticut*, 419–28. Agricultural Experiment Station,

Eduardo, Iban, Pere Arús, Antonio José Monforte, Javier Obando, Juan Pablo Fernández-Trujillo, Juan Antonio Martínez, Antonio Luís Alarcón, Jose María María Álvarez, and Esther van der Knaap. 2007. "Estimating the Genetic Architecture of Fruit Quality Traits in Melon Using a Genomic Library of near Isogenic Lines." *Journal of the American Society for Horticultural Science* 132 (1): 80–9. https://doi.org/10.21273/JASHS.132.1.80.

Esmaeili, M., F. Soltani, M. R. Bihamta, and M. Javan Nikkhah. 2022. "Estimation of Yield Combining Ability and Fruit-Related Traits Using Diallel Analysis in Melon (*Cucumis melo* L.)." *International Journal of Horticultural Science & Technology* 9 (1): 131–42.

FAOSTAT [Database]. 2025. *Production*. Accessed May 9, 2025, https://www.fao.org/faostat/en/#data/QCL/visualize.

Feyzian, E., H. Dehghani, A. M. Rezai, and M. J. Jalali Javaran. 2009. "Diallel Cross Analysis for Maturity and Yield-Related Traits in Melon (*Cucumis melo* L.)." *Euphytica* 168 (2): 215–23. https://doi.org/10.1007/s10681-009-9904-9.

Galpaz, Navot, Itay Gonda, Doron Shem-Tov, Omer Barad, Galil Tzuri, Shery Lev, Zhangjun Fei, Yimin Xu, Linyong Mao, Chen Jiao, Rotem Harel-Beja, Adi Doron-Faigenboim, Oren Tzfadia, Einat Bar, Ayala Meir, Uzi Sa'ar, Aaron Fait, Eran Halperin, Merav Kenigswald, Elazar Fallik, Nadia Lombardi, Guy Kol, Gil Ronen, Yosef Burger, Amit Gur, Ya'akov Tadmor, Vitaly Portnoy, Arthur A. Schaffer, Efraim Lewinsohn, James J. Giovannoni, and Nurit Katzir. 2018. "Deciphering Genetic Factors That Determine Melon Fruit-Quality Traits Using RNA-Seq-Based High-Resolution QTL and eQTL Mapping." *Plant Journal: For Cell & Molecular Biology* 94 (1): 169–91. https://doi.org/10.1111/tpj.13838.

Garcia-Mas, Jordi, Andrej Benjak, Walter Sanseverino, Michael Bourgeois, Gisela Mir, Víctor M. González, Elizabeth Hénaff, Francisco Câmara, Luca Cozzuto, Ernesto Lowy, Tyler Alioto, Salvador Capella-Gutiérrez, Jose Blanca, Joaquín Cañizares, Pello Ziarsolo, Daniel Gonzalez-Ibeas, Luis Rodríguez-Moreno, Marcus Droege, Lei Du, Miguel Alvarez-Tejado, Belen Lorente-Galdos, Marta Melé, Luming Yang, Yiqun Weng, Arcadi Navarro, Tomas Marques-Bonet, Miguel A. Aranda, Fernando Nuez, Belén Picó, Toni Gabaldón, Guglielmo Roma, Roderic Guigó, Josep M. Casacuberta, Pere Arús, and Pere Puigdomènech. 2012. "The Genome of Melon (*Cucumis melo* L.)." *Proceedings of the National Academy of Sciences of the United States of America* 109 (29): 11872–7. https://doi.org/10.1073/pnas.1205415109.

Gerland, Patrick, Adrian E. Raftery, Hana Sevčíková, Nan Li, Danan Gu, Thomas Spoorenberg, Leontine Alkema, Bailey K. Fosdick, Jennifer Chunn, Nevena Lalic, Guiomar Bay, Thomas Buettner, Gerhard K. Heilig, and John Wilmoth. 2014. "World Population Stabilization Unlikely This Century." *Science* 346 (6206): 234–7. https://doi.org/10.1126/science.1257469.

Goldman, Amy. 2002. "Melons for the passionate Grower, Artisan, New York Conditions." *Mus. Kem. Üniv. Tar. Bili. Der.* 27: 309–17.

Harel-Beja, R., G. Tzuri, V. Portnoy, M. Lotan-Pompan, S. Lev, S. Cohen, N. Dai, L. Yeselson, A. Meir, S. E. Libhaber, E. Avisar, T. Melame, P. van Koert, H. Verbakel, R. Hofstede, H. Volpin, M. Oliver, A. Fougedoire, C. Stalh, J. Fauve, B. Copes, Z. Fei, J. Giovannoni, N. Ori, E. Lewinsohn, A. Sherman, J. Burger, Y. Tadmor, A. A. Schaffer, and N. Katzir. 2010. "A Genetic Map of Melon Highly Enriched with Fruit Quality QTLs and EST Markers, Including Sugar and Carotenoid Metabolism Genes." *Theoretical & Applied Genetics* 121 (3): 511–33. https://doi.org/10.1007/s00122-010-1327-4.

Heckenberger, M., M. Bohn, and A. E. Melchinger. 2005. "Identification of Essentially Derived Varieties Obtained from Biparental Crosses of Homozygous Lines: I. Simple Sequence Repeat Data from Maize Inbreds." *Crop Science* 45 (3): 1120–31. https://doi.org/10.2135/cropsci2004.0110.

Huang, Xuehui, Shihua Yang, Junyi Gong, Qiang Zhao, Q. Feng, Qilin Zhan, Yan Zhao, Wenjun Li, Benyi Cheng, Junhui Xia, Neng Chen, Tao Huang, Lei Zhang, Danlin Fan, Jiaying Chen, Congcong Zhou, Yiqi Lu, Qijun Weng, and Bin Han. 2016. "Genomic Architecture of Heterosis for Yield Traits in Rice." *Nature* 537 (7622): 629–33. https://doi.org/10.1038/nature19760.

José, Monforte Antonio, Eduardo Iban, Abad Silvia, and Arús Pere. 2005. "Inheritance Mode of Fruit Traits in Melon: Heterosis for Fruit Shape and Its Correlation with Genetic Distance." *Euphytica* 144 (1–2): 31–8. https://doi.org/10.1007/s10681-005-0201-y.

Kaeppler, Shawn. 2012. "Heterosis: Many Genes, Many Mechanisms – End the Search for an Undiscovered Unifying Theory." *ISRN Botany* 2012: 1–12. https://doi.org/10.5402/2012/682824.

Kalb, Thomas J., and D. W. Davis. 1984. "Evaluation of Combining Ability, Heterosis, and Genetic Variance for Yield, Maturity, and Plant Characteristics in Bush Muskmelon." *Journal of the American Society for Horticultural Science* 109 (3): 416–9. https://doi.org/10.21273/JASHS.109.3.416.

Kaushik, Prashant, Mariola Plazas, Jaime Prohens, Santiago Vilanova, and Pietro Gramazio. 2018. "Diallel Genetic Analysis for Multiple Traits in Eggplant and Assessment of Genetic Distances for Predicting Hybrids Performance." *PLoS One* 13 (6): e0199943. https://doi.org/10.1371/journal.pone.0199943.

Kesh, Hari, and Prashant Kaushik. 2021. "Advances in Melon (*Cucumis melo* L.) Breeding: An Update." *Scientia Horticulturae* 282. https://doi.org/10.1016/j.scienta.2021.110045.

Li, Dayong, Zhiyuan Huang, Shuhui Song, Yeyun Xin, Donghai Mao, Qiming Lv, Ming Zhou, Dongmei Tian, Mingfeng Tang, Q. Wu, Xue Liu, Tingting Chen, Xianwei Song, Xiqin Fu, Bingran Zhao, Chengzhi Liang, Aihong Li, Guozhen Liu, Shigui Li, Songnian Hu, Xiaofeng Cao, Jun Yu, Longping Yuan, Caiyan Chen, and Lihuang Zhu. 2016. "Integrated Analysis of Phenome, Genome, and Transcriptome of Hybrid Rice Uncovered Multiple Heterosis-Related Loci for Yield Increase." *Proceedings of the National Academy of Sciences of the United States of America* 113 (41): E6026–35. https://doi.org/10.1073/pnas.1610115113.

Lian, Qun, Qiushi Fu, Yongyang Xu, Zhicheng Hu, Jing Zheng, Aiai Zhang, Yuhua He, Changsheng Wang, Chuanqiang Xu, Benxue Chen, Jordi Garcia-Mas, Guangwei Zhao, and Huaisong Wang. 2021. "QTLs and Candidate Genes Analyses for Fruit Size under Domestication and Differentiation in Melon (*Cucumis melo* L.) Based on High Resolution Maps." *BMC Plant Biology* 21 (1): 126. https://doi.org/10.1186/s12870-021-02904-y.

Lippert, L. F., and P. D. Legg. 1972. "Diallel Analyses for Yield and Maturity Characteristics in Muskmelon Cultivars." *Journal of the American Society for Horticultural Science* 97 (1): 87–90. https://doi.org/10.21273/JASHS.97.1.87.

Liu, Shi, Peng Gao, Qianglong Zhu, Zicheng Zhu, Hongyu Liu, Xuezheng Wang, Yiqun Weng, Meiling Gao, and Feishi Luan. 2020. "Resequencing of 297 Melon Accessions Reveals the Genomic History of Improvement and Loci Related to Fruit Traits in Melon." *Plant Biotechnology Journal* 18 (12): 2545–58. https://doi.org/10.1111/pbi.13434.

Luan, Feishi, Yunyan Sheng, Yuhan Wang, and Jack E. Staub. 2010. "Performance of Melon Hybrids Derived from Parents of Diverse Geographic Origins." *Euphytica* 173 (1): 1–16. https://doi.org/10.1007/s10681-009-0110-6.

Mayobre, Carlos, Lara Pereira, Abdelali Eltahiri, Einat Bar, Efraim Lewinsohn, Jordi Garcia-Mas, and Marta Pujol. 2021. "Genetic Dissection of Aroma Biosynthesis in Melon and Its Relationship with Climacteric Ripening." *Food Chemistry* 353: 129484. https://doi.org/10.1016/j.foodchem.2021.129484.

Melchinger, A. E. 1999. "Genetic Diversity and Heterosis." In *Genetics and Exploitation of Heterosis in Crops*, edited by James G. Coors and Shivaji Pandey, 99–118. Madison, WI, US: American Society of Agronomy, Crop Science Society of America, https://doi.org/10.2134/1999.geneticsandexploitation.c10.

Melo, D. R. M. et al. 2011. "Controle Genético da Herança Sólidos Solúveis em Frutos de Meloeiro." *Journal of the Interamerican Society for Tropical Horticulture* 54: 146–50.

Monforte Antonio J., Eduardo Iban, Abad Silvia, and Arús Pere. 2005. "Inheritance Mode of Fruit Traits in Melon: Heterosis for Fruit Shape and Its Correlation with Genetic Distance." *Euphytica* 144 (1–2): 31–8. https://doi.org/10.1007/s10681-005-0201-y.

Monforte, Antonio J., M. Oliver, M. J. Gonzalo, J. M. Alvarez, R. Dolcet-Sanjuan, and P. Arús. 2004. "Identification of Quantitative Trait Loci Involved in Fruit Quality Traits in Melon (*Cucumis melo* L.)." *Theoretical & Applied Genetics* 108 (4): 750–8. https://doi.org/10.1007/s00122-003-1483-x.

Moreno, Eduard, Javier M. Obando, Noelia Dos-Santos, J. Pablo Fernández-Trujillo, Antonio J. Monforte, and Jordi Garcia-Mas. 2008. "Candidate Genes and QTLs for Fruit Ripening and Softening in Melon." *Theoretical & Applied Genetics* 116 (4): 589–602. https://doi.org/10.1007/s00122-007-0694-y.

Munger, H. M. 1942. "The Possible Utilization of First Generation Muskmelon Hybrids and Improved Methods of Hybridization." *Proceedings of the American Society for Horticultural Science* 40: 405–10.

Napolitano, Marco. 2015. *Exploring Heterosis in melon (Cucumis melo L.)* [PhD Thesis]. Perugia, Italy: Università degli Studi di Perugia.

Napolitano, Marco, Niccolò Terzaroli, Subash Kashyap, Luigi Russi, Elen Jones-Evans, and Emidio Albertini. 2020. "Exploring Heterosis in Melon (*Cucumis melo* L.)." *Plants* 9 (2): 282. https://doi.org/10.3390/plants9020282.

Nimmakayala, Padma, Yan R. Tomason, Venkata L. Abburi, Alejandra Alvarado, Thangasamy Saminathan, Venkata G. Vajja, Germania Salazar, Girish K. Panicker, Amnon Levi, William P. Wechter, James D. McCreight, Abraham B. Korol, Yefim Ronin, Jordi Garcia-Mas, and Umesh K. Reddy. 2016. "Genome-Wide Differentiation of Various Melon Horticultural Groups for Use in GWAS for Fruit Firmness and Construction of a High Resolution Genetic Map." *Frontiers in Plant Science* 7: 1437. https://doi.org/10.3389/fpls.2016.01437.

Paris, Miriam K., Juan E. Zalapa, James D. McCreight, and Jack E. Staub. 2008. "Genetic Dissection of Fruit Quality Components in Melon (*Cucumis melo* L.) Using a RIL Population Derived from Exotic × Elite US Western Shipping Germplasm." *Molecular Breeding* 22 (3): 405–19. https://doi.org/10.1007/s11032-008-9185-3.

Pereira, L., V. Ruggieri, S. Pérez, K. G. Alexiou, M. Fernández, T. Jahrmann, M. Pujol, and J. Garcia-Mas. 2018. "QTL Mapping of Melon Fruit Quality Traits Using a High-Density GBS-Based Genetic Map." *BMC Plant Biology* 18 (1): 324. https://doi.org/10.1186/s12870-018-1537-5.

Perpiñá, G., S. Roselló, C. Esteras, J. Beltrán, A. J. Monforte, J. Cebolla-Cornejo, and B. Picó. 2021. "Analysis of Aroma-Related Volatile Compounds Affected by 'Ginsen Makuwa' Genomic Regions Introgressed in 'Vedrantais' Melon Background." *Scientia Horticulturae* 276: 109664. https://doi.org/10.1016/j.scienta.2020.109664.

Pouyesh, Alireza, Mahmoud Lotfi, Hossein Ramshini, Ezzat Karami, Aliasghar Shamsitabar, and Elahe Armiyoun. 2017. "Genetic Analysis of Yield and Fruit Traits in Cantaloupe Cultivars." *Plant Breeding* 136 (4): 569–77. https://doi.org/10.1111/pbr.12486.

Rad, M. N., M. Allahdoo, and H. R. Fanaei. 2010. "Study of Some Yield Traits Relationship in Melon (*Cucumis melo* L.) Germplasm Gene Bank of Iran by Correlation and Factor Analysis." *Trakia Journal of Sciences* 8: 27–32.

Ramamurthy, Raghuprakash Kastoori, and Brian M. Waters. 2015. "Identification of Fruit Quality and Morphology QTLs in Melon (*Cucumis melo*) Using a Population Derived from *flexuosus* and *Cantalupensis* Botanical Groups." *Euphytica* 204 (1): 163–77. https://doi.org/10.1007/s10681-015-1361-z.

Schemberger, Michelle O., Marília A. Stroka, Letícia Reis, Kamila Karoline de Souza Los, Gillize Aparecida Telles de Araujo, Michelle Zibetti Tadra Sfeir, Carolina Weigert Galvão, Rafael Mazer Etto, Amanda Regina Godoy Baptistão, and Ricardo Antonio Ayub. 2020. "Transcriptome Profiling of Non-climacteric 'Yellow' Melon during Ripening: Insights on Sugar Metabolism." *BMC Genomics* 21 (1): 262. https://doi.org/10.1186/s12864-020-6667-0.

Seymour, Danelle K., Eunyoung Chae, Dominik G. Grimm, Carmen Martín Martín Pizarro, Anette Habring-Müller, François Vasseur, Barbara Rakitsch, Karsten M. Borgwardt, Daniel Koenig, and Detlef Weigel. 2016. "Genetic Architecture of Nonadditive Inheritance in *Arabidopsis thaliana* Hybrids." *Proceedings of the National Academy of Sciences of the United States of America* 113 (46): E7317–26. https://doi.org/10.1073/pnas.1615268113.

Shahwar, Durre, Zeba Khan, and Younghoon Park. 2023. "Molecular Marker-Assisted Mapping, Candidate Gene Identification, and Breeding in Melon (*Cucumis melo* L.): A Review." *International Journal of Molecular Sciences* 24 (20): 15490. https://doi.org/10.3390/ijms242015490.

Shajari, Moein, Forouzandeh Soltani, Mohammad Reza Bihamta, and Michael Alabboud. 2021. "Genetic Analysis and Inheritance of Floral and Fruit Traits in Melon (*Cucumis melo*) in the Full Diallel Cross." *Plant Breeding* 140 (3): 486–96. https://doi.org/10.1111/pbr.12915.

Shashikumar, K. T., and M. Pitchaimuthu. 2016. "Heterosis and Combining Ability Analysis of Quantitative and Qualitative Traits in Muskmelon (*Cucumis melo* L.)." *International Journal of Agricultural Science & Research* 6: 341–8.

Shattuck, V. I., B. Christie, and C. Corso. 1993. "Principles for Griffing's Combining Ability Analysis." *Genetica* 90 (1): 73–7. https://doi.org/10.1007/BF01435180.

Singh, M. J., and K. S. Randhawa. 1990. "Evaluation of Heterosis and Ability for Traits in Muskmelon." *Indian Journal of Horticulture* 47: 228–32.

Tang, Yufan, Chong Zhang, Songxiao Cao, Xiao Wang, and Hongyan Qi. 2015. "The Effect of CmLOXs on the Production of Volatile Organic Compounds in Four Aroma Types of Melon (Cucumis melo)." *PLoS One* 10 (11): e0143567. https://doi.org/10.1371/journal.pone.0143567.

Teklewold, Adefris, and Heiko C. Becker. 2006. "Comparison of Phenotypic and Molecular Distances to Predict Heterosis and F1 Performance in Ethiopian Mustard (Brassica carinata A. Braun)." *Theoretical & Applied Genetics* 112 (4): 752–9. https://doi.org/10.1007/s00122-005-0180-3.

Tomkowiak, Agnieszka, Jan Bocianowski, Dominika Radzikowska, and Przemysław Łukasz Kowalczewski. 2019. "Selection of Parental Material to Maximize Heterosis Using SNP and SilicoDarT Markers in Maize." *Plants* 8 (9): 349. https://doi.org/10.3390/plants8090349.

Troyer, A. Forrest. 2006. "Adaptedness and Heterosis in Corn and Mule Hybrids." *Crop Science* 46 (2): 528–43. https://doi.org/10.2135/cropsci2005.0065.

Tzuri, Galil, Asaf Dafna, Ben Itzhaki, Ilan Halperin, Elad Oren, Tal Isaacson, Adi Faigenboim, Yelena Yeselson, Harry S. Paris, Michael Mazourek, Joseph Burger, Arthur A. Schaffer, and Amit Gur. 2025. "Meta Genetic Analysis of Melon Sweetness." *Theoretical & Applied Genetics* 138 (4): 68. https://doi.org/10.1007/s00122-025-04863-6.

Valério, I. P., F. I. F. Carvalho, A. C. Oliveira, V. Q. Souza, G. Benin, D. A. M. Schmidt, G. Ribeiro, R. Nornberg, and H. Luch. 2009. "Combining Ability of Wheat Genotypes in Two Models of Diallel Analyses." *Cropp Breeding & Applied Biotechnology* 9 (2): 100–7. https://doi.org/10.12702/1984-7033.v09n02a01.

Wang, Yu-Hua, Dong-Hong Wu, Jin-Hsing Huang, Shing-Jy. Tsao, Kae-Kang Hwu, and Hsiao-Feng Lo. 2016. "Mapping Quantitative Trait Loci for Fruit Traits and Powdery Mildew Resistance in Melon (*Cucumis melo*)." *Botanical Studies* 57 (1): 19. https://doi.org/10.1186/s40529-016-0130-1.

Wheeler, Tim, and Joachim von Braun. 2013. "Climate Change Impacts on Global Food Security." *Science* 341 (6145): 508–13. https://doi.org/10.1126/science.1239402.

Xu, Longlan, Yuhua He, Lingli Tang, Yongyang Xu, and Guangwei Zhao. 2022. "Genetics, Genomics, and Breeding in Melon." *Agronomy* 12 (11): 2891. https://doi.org/10.3390/agronomy12112891.

Yan, W. 1999. *Methodology of Cultivar Evaluation Based on Yield Trial Data – With Special Reference to Winter Wheat in Ontario* [PhD Thesis]. Guelph, ON, Canada: University of Guelph.

Yan, Weikai, and L. A. Hunt. 2002. "Biplot Analysis of Diallel Data." *Crop Science* 42 (1): 21–30. https://doi.org/10.2135/cropsci2002.0021.

Yan, Weikai, L. A. Hunt, Qinglai Sheng, and Zorka Szlavnics. 2000. "Cultivar Evaluation and Mega-environment Investigation Based on the GGE Biplot." *Crop Science* 40 (3): 597–605. https://doi.org/10.2135/cropsci2000.403597x.

Yang, Jinliang, Sofiane Mezmouk, Andy Baumgarten, Edward S. Buckler, Katherine E. Guill, Michael D. McMullen, Rita H. Mumm, and Jeffrey Ross-Ibarra. 2017. "Incomplete Dominance of Deleterious Alleles Contributes Substantially to Trait Variation and Heterosis in Maize." *PLoS Genetics* 13 (9): e1007019. https://doi.org/10.1371/journal.pgen.1007019.

Zalapa, J. E., J. E. Staub, and J. D. McCreight. 2006. "Generation Means Analysis of Plant Architectural Traits and Fruit Yield in Melon." *Plant Breeding* 125 (5): 482–7. https://doi.org/10.1111/j.1439-0523.2006.01273.x.

Zalapa, J. E., J. E. Staub, J. D. McCreight, S. M. Chung, and H. Cuevas. 2007. "Detection of QTL for Yield-Related Traits Using Recombinant Inbred Lines Derived from Exotic and Elite US Western Shipping Melon Germplasm." *Theoretical & Applied Genetics* 114 (7): 1185–201. https://doi.org/10.1007/s00122-007-0510-8.

Zhang, Chong, Yazhong Jin, Jieying Liu, Yufan Tang, Songxiao Cao, and Hongyan Qi. 2014. "The Phylogeny and Expression Profiles of the Lipoxygenase (LOX) Family Genes in the Melon (*Cucumis melo* L.) Genome." *Scientia Horticulturae* 170: 94–102. https://doi.org/10.1016/j.scienta.2014.03.005.

Zhang, Hong, Hongping Yi, Mingzhu Wu, Yongbin Zhang, Xuejin Zhang, Meihua Li, and Guangzhi Wang. 2016. "Mapping the Flavor Contributing Traits On "Fengwei Melon" (*Cucumis melo* L.) Chromosomes Using Parent Resequencing and Super Bulked-Segregant Analysis." *PLoS One* 11 (2): e0148150. https://doi.org/10.1371/journal.pone.0148150.

Zhang, Taifeng, Zhuo Ding, Jiajun Liu, Boyan Qiu, and Peng Gao. 2020. "QTL Mapping of Pericarp and Fruit-Related Traits in Melon (*Cucumis melo* L.) Using SNP-Derived CAPS Markers." *Scientia Horticulturae* 265: 109243. https://doi.org/10.1016/j.scienta.2020.109243.

Index

Note: Page numbers in *italics* indicate a figure and page numbers in **bold** indicate a table on the corresponding page.

E

F

G

H

I

J

K

L

M

N

P

Q

R

S

T

V

W

Y

Z